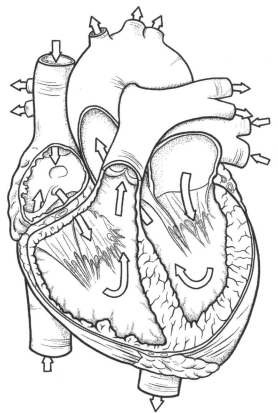

THE **HUMAN BODY**
COLORING BOOK
FROM CELLS TO SYSTEMS AND BEYOND

Ellyn Elson
Lawrence M. Elson, PhD

Collins Reference

An Imprint of HarperCollins*Publishers*

WRITTEN BY:
Ellyn Elson
Lawrence M. Elson, PhD

EDITOR:
Andrea Good

ART DIRECTOR:
Andrea Good

ILLUSTRATORS:
Allison Anderson
Jessica Fraser
Andrea Good
Annie Kendall
Dana Longuevan

This coloring book was adapted from *It's Your Body* by Lawrence Elson, originally published in 1975 by McGraw Hill.

THE HUMAN BODY COLORING BOOK. Copyright © 2022 by Coloring Concepts, Inc. All rights reserved. Printed in Malaysia. No part of this book may be used or reproduced in any manner whatsoever without written permission except in the case of brief quotations embodied in critical articles and reviews. For information, address HarperCollins Publishers, 195 Broadway, New York, NY 10007.

HarperCollins books may be purchased for educational, business, or sales promotional use. For information, please email the Special Markets Department at SPsales@harpercollins.com.

This book was produced by Coloring Concepts Inc.,123 Sasha Court, Napa, CA 94558.

Library of Congress Cataloging-in-Publication Data has been applied for.

ISBN 978-0-06-300975-2

22 23 24 25 26   10 9 8 7 6 5 4 3 2 1

CONTENTS

CONTENTS

This book is dedicated to all individuals who want to better understand what makes this miraculous machine, the *body*, work so elegantly.

Learning about the intricacies and design of the human body has been important to mankind since humans have learned language and how to communicate. Our curiosity about how our body functions continues for a lifetime. The original *It's Your Body* textbook, published in 1975, has satisfied our need to know for over forty years.

After using *It's Your Body* as a reference text for the past twenty years, I always felt that it would have been better as a coloring book rather than a hardcover textbook. When the opportunity presented itself to add new titles to our library of books, re-creating *It's Your Body* was on the top of the list. Thus, *The Human Body Coloring Book* was born.

Redesigning the textbook into a coloring book allowed us to use the method of **kinesthetic learning**, where physical interaction with a subject facilitates learning. The involvement of body movement can help students and inquisitive learners better understand difficult academic content.

I've never forgotten my first course in human anatomy. Even with my strong background in nutrition and anatomy, it was a difficult class. If there had been a coloring book to help me better understand what makes the body fuction, it would have made studying much more meaningful. Whether you are taking your first course in human anatomy or simply want to learn more about the human body, I hope this coloring book will bring you hours of enjoyment.

Many thanks to Larry Elson for his encouragement and expertise and to Andrea Good, the most creative and diligent artist that I have ever had the opportunity to work with, for bringing this book to market. A special acknowledgment to Russell Peterson, the original illustrator, who set a high standard for many of the illustrations that were re-created in this book.

Happy coloring!

Ellyn

Ellyn Elson

HELPFUL HINTS FOR USING THIS BOOK:

To facilitate learning, words in **bold** indicate important terms in both the text and illustrations.

Some labeled components of the illustrations are not explained in the text. Please use your own resources to further explore words or terms you are curious about!

It will be helpful to follow specific coloring instructions where provided to more easily understand the subject matter. Colored pencils are best to use in this book as crayons might make it difficult to capture the smaller details, and markers may bleed through the page.

SECTION ONE:
THE ORGANIZATION OF YOUR BODY

The human body is made up of a large number of structures that may be ranked from complex to relatively simple. The complex structure in this arrangement might be the brain. A simple structure might be fat cells in connective tissue. It is true that you cannot live without a brain but you cannot live without cells storing fat, either. Which is more important?

Neither. They are interdependent. Each is as important as the other.

Although there is an ordered system of structure, from large to small, from complex to simple, in general no one part of the system is more important than another. The individual parts are functionally and frequently structurally interrelated in order for the whole body to function as a living organism.

The classification of structure may have a *functional* or a *regional* (location) basis. In this coloring book we will take a regional look at the body, to allow the reader to understand the body systemically and dissected regionally.

Beginning with cells, the simplest living units of the body, one finds the smaller and less complex structures are either *components* of the cells or *products* of the cells, while larger and more complex structures are often *aggregations* of cells and their products (supporting fibers), infused in a matrix or ground substance (a product of cells). For example, note the constitution of the brain:

Nerve cells and processes (neurons)	+	supporting cells and processes (neuroglia)	+	connective tissue cells and fibers with ground substance

TISSUES: Most cells of the body are neither isolated nor randomly placed. Instead, they are congregated into units having a related function. These aggregations of cells, often attached by intercellular "cement" or by fibers, are termed *tissues*. The whole body is structured from four basic tissues: epithelial, connective, muscle, and nervous.

ORGANS: The four fundamental tissues are structurally and functionally interrelated to such a degree that they form a complex or structure in themselves. Such a structure is called an *organ*. Examples of organs are arteries, veins, skin, glands, nerves, liver, kidneys, gallbladder, heart, lungs, brain, etc.

SYSTEMS: In most cases, certain organs are associated together in performing an overall function. Those organs that are associated with a common function constitute a *system*. The eleven systems of the body include all of the cells, tissues, and organs. They are the integumentary, digestive, cardiovascular, respiratory, reproductive, muscular, lymphatic, nervous, urinary, skeletal, and endocrine systems.

TERMS OF POSITION & DIRECTION

When looking at the body from a regional viewpoint, it may be noted that the three main regions of the body are the head, trunk, and limbs, and within each of those areas, we find subregions:

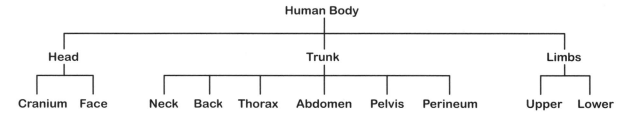

Let us now familiarize ourselves with the body in terms of position and direction, for reference throughout this book. When describing any body location, orientation, direction, or movement, the standard "map" is the **anatomical position**: The body is erect and the arms are at the sides with the palms facing toward the front. The lower limbs are upright and parallel with the feet side-by-side and the toes directed forward.

Listed below are the terms of position and direction, and it is important that you commit them to memory.

Cranial, superior: refers to the head.
Anterior, ventral: pertains to the front of the body: chest, abdomen, shins, palms.
Rostral: toward the oral or nasal region of the head.
Posterior, dorsal: pertains to the upper side or back of the body.
Medial: pertaining to the middle or toward the middle of the body.
Lateral: away from the midline of the body; side or part that is away from the center of the body.
Proximal: closer to the body trunk or the body center.
Distal: away from; the more (or most) distant of two (or more) things; opposite of proximal.
Caudal, inferior: away from the head.
Superficial: on the surface or shallow.
Deep: away from the surface or further into the body.
Ipsilateral: on the same side.
Contralateral: on the opposite side.

Assuming that the body is upright and in the anatomical position:

• The nose is **anterior** to the back of the head.
• The shoulder is **lateral** to and **inferior** to the neck.
• The knee joint is **superior** to the ankle joint.
• The layer of fat just **deep** to the skin is said to be **superficial** to a deeper layer of connective or muscle tissue.

INSTRUCTIONS FOR COLORING LABELED ILLUSTRATIONS: Use the same color for each arrow and related label on the illustration. For example, color the arrow **A** and the words CRANIAL, SUPERIOR all with the same color. Get creative and have fun!

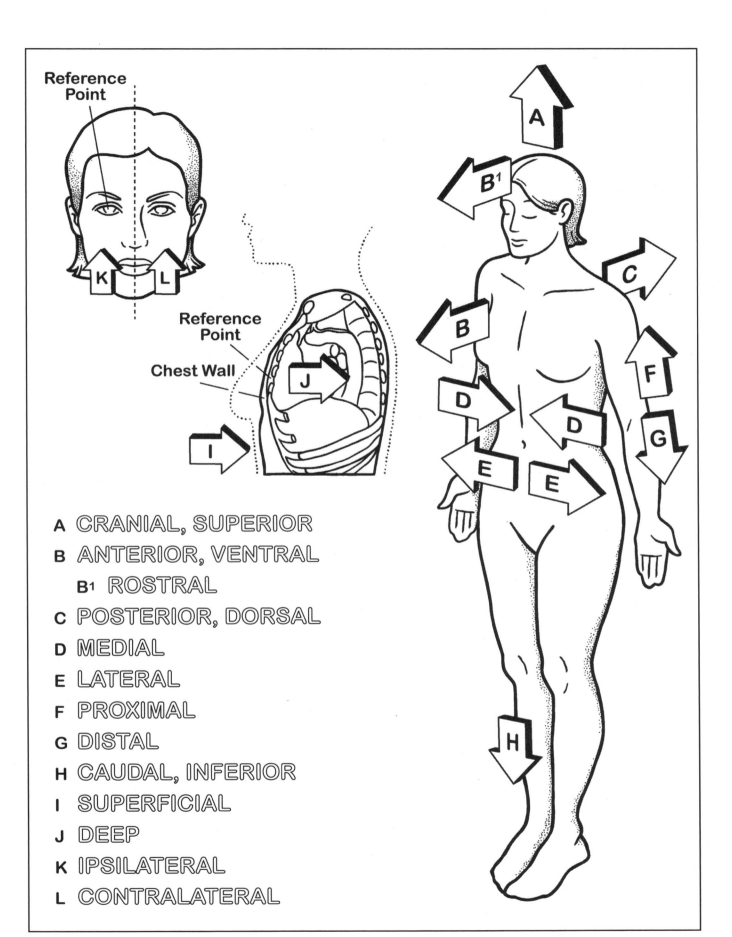

Reference Point

Reference Point

Chest Wall

A CRANIAL, SUPERIOR
B ANTERIOR, VENTRAL
   B¹ ROSTRAL
C POSTERIOR, DORSAL
D MEDIAL
E LATERAL
F PROXIMAL
G DISTAL
H CAUDAL, INFERIOR
I SUPERFICIAL
J DEEP
K IPSILATERAL
L CONTRALATERAL

ANATOMIC PLANES & SECTIONS

To describe the views from which you observe the body and to understand its structural arrangement, it is helpful to study the body by means of the **anatomic planes** and **sections**.

Median (**midsagittal**) **plane**: Relative to the whole body, this is also the longitudinal (vertical) axis. It divides the body into equal left and right halves.

Sagittal plane: A longitudinal axis on either side of the median plane, dividing the body into unequal left and right parts.

Coronal or **frontal plane**: Divides the body into anterior (in front of) and posterior (in back of) parts.

Transverse plane or **cross section**: Relative to the whole body, this is also the horizontal plane.

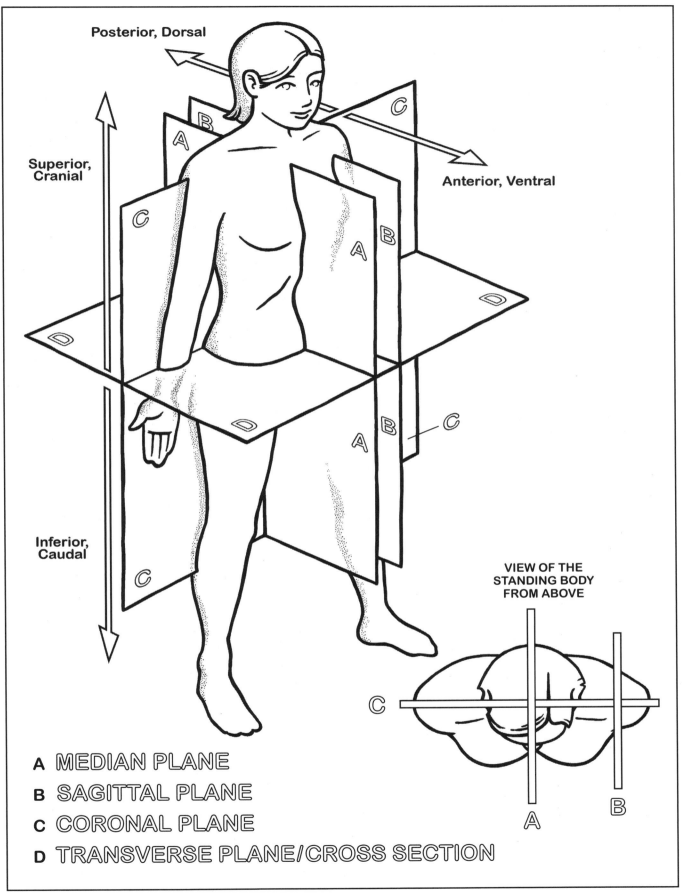

Posterior, Dorsal

Superior,
Cranial

Anterior, Ventral

Inferior,
Caudal

VIEW OF THE
STANDING BODY
FROM ABOVE

A MEDIAN PLANE
B SAGITTAL PLANE
C CORONAL PLANE
D TRANSVERSE PLANE/CROSS SECTION

CAVITIES & LININGS: CLOSED BODY CAVITIES

CLOSED CAVITIES

The body is organized so that the various **viscera** (the internal organs of the main cavity of the body) occupy closed spaces called **cavities**. Those cavities behind the midline frontal plane of the body (separating the body into equal front and back halves) are the **posterior cavities** and those in front are the **anterior cavities**.

The posterior cavities, from the head down, are the **cranial cavity**, housing the **brain**, and **vertebral cavity**, containing the **spinal cord**. Both are surrounded and protected by the **dura mater**, a layer of connective tissue.

The anterior cavities from top to bottom are the **thoracic cavity**, including the **pleural cavity** (lungs), the **pericardial cavity** (heart), and the **abdominopelvic cavity**.

SEROUS MEMBRANES

The membranes that line the pleural, pericardial, and peritoneal cavities are known as **serous membranes**. Each serous membrane consists of a single layer of cells and supporting tissue. These cells secrete a watery **serous** fluid that prevents friction when two such membranes rub against one another, as they are prone to do with organ movement.

ABDOMINOPELVIC CAVITY

The abdominopelvic cavity includes the cavity of the **abdomen** and **pelvis**. The roof of this cavity is the **thoracic diaphragm**. The floor is a muscle, called, with its coverings of tissue, the **pelvic diaphragm**. This cavity houses the alimentary tract, liver, gallbladder, pancreas, spleen, urinary organs, and the internal organs of reproduction, as well as innumerable ducts, vessels, and nerves.

Many of these organs are enclosed by the **peritoneum**, a protective membrane lining the inside walls of the abdominal cavity. It consists of two layers:

• The **parietal peritoneum**, an outer layer that lines the abdominal and pelvic cavities
• The **visceral peritoneum**, an inner layer that covers the surfaces of most abdominal organs

The **peritoneal cavity** is the space between the parietal and visceral layers of the peritoneum. It is known as a **potential space**: in anatomical terminology, the area between two adjacent structures that are usually pressed together. This space is "potential" rather than realized, like a balloon that has no interior volume until inflated.

PERITONEUM

1 PARIETAL PERITONEUM
2 VISCERAL PERITONEUM
3 PERITONEAL CAVITY

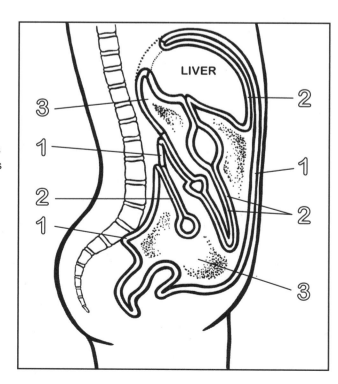

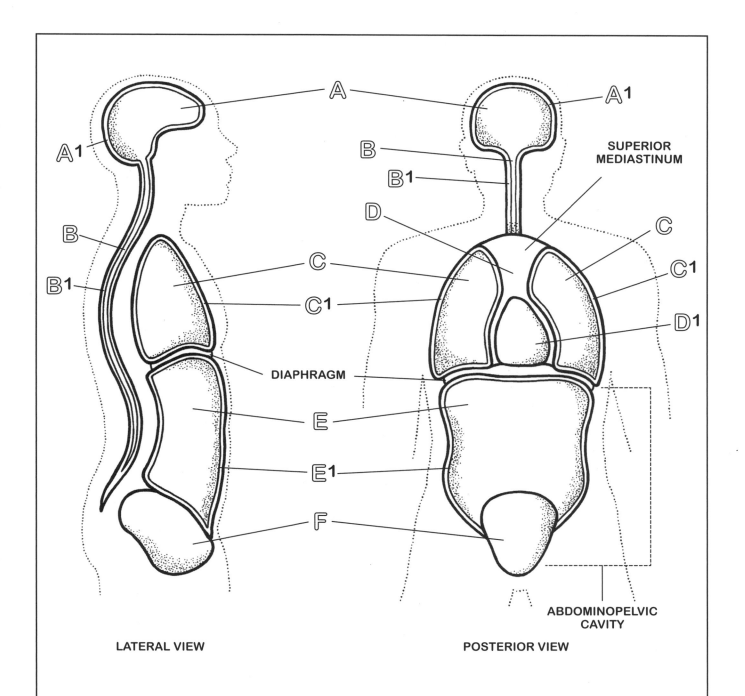

LATERAL VIEW

POSTERIOR VIEW

SUPERIOR MEDIASTINUM

DIAPHRAGM

ABDOMINOPELVIC CAVITY

A CRANIAL CAVITY
   A1 DURA MATER
B VERTEBRAL CAVITY
   B1 DURA MATER
C THORACIC CAVITY
   C1 PLEURA

D MEDIASTINUM
   D1 PERICARDIAL CAVITY
E ABDOMINAL CAVITY
   E1 PERITONEUM
F PELVIC CAVITY

CAVITIES & LININGS: OPEN VISCERAL CAVITIES

OPEN CAVITIES AND MUCOUS MEMBRANES

Many organs of the body are held and protected within **open visceral cavities**: fluid-filled spaces that can also process, store, and transport materials, either ingested or inhaled from outside (as in the case of the alimentary or respiratory tracts) or secreted from inside (as in the case of the reproductive and urinary tracts). These cavities are, at one or both ends, continuous with the outside of your body.

Open visceral cavities are lined with a multilayered **mucous membrane**. Mucous membranes (mucosa) contain organized strata of lining cells (epithelia), connective tissue, and frequently muscle tissue. Mucous membranes supply the labor force for the tasks of the organ. Epithelial cells do the processing (secretion of enzymes, absorption, etc.), and smooth muscle provides for the movement and mixing of materials in the cavity, or lumen (opening or interior of a vessel). The mucosa also serves to keep the lumen moist. The mucosa is supported by an underlying connective tissue layer (submucosa) and a thick layer, or layers, of muscle.

The kidneys, ureters, and urinary bladder lie **deep** (behind) to the posterior peritoneum of the abdominal cavity (see page 7). In other words, they are **retroperitoneal**.

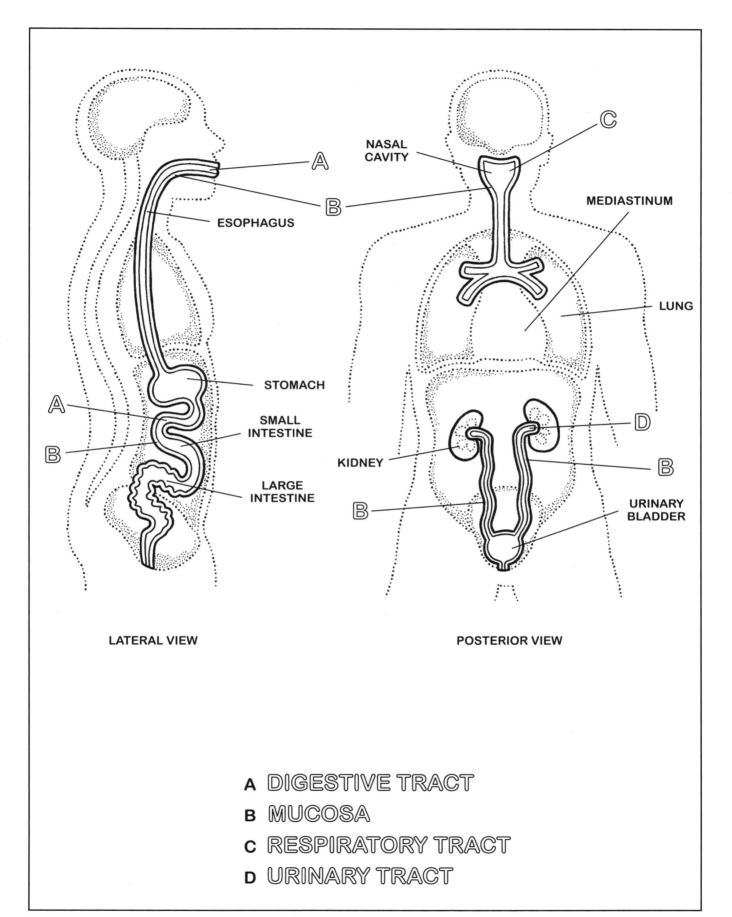

A   NASAL CAVITY

ESOPHAGUS

B

C   MEDIASTINUM

LUNG

STOMACH

SMALL INTESTINE

A

B   KIDNEY

LARGE INTESTINE

D

B

URINARY BLADDER

LATERAL VIEW

POSTERIOR VIEW

A DIGESTIVE TRACT
B MUCOSA
C RESPIRATORY TRACT
D URINARY TRACT

MAJOR VISCERAL ORGANS OF THE BODY

You have learned that cavities accommodate organs and other structures; familiarize yourself with the organs these cavities house and protect. All humans have vital organs that are essential for survival. Each organ has a specific role that contributes to the overall well-being of the human body. A group of organs whose jobs are closely related are often referred to as a **system**.

BRAIN
The **brain** is the control center of the nervous system, located within the skull. Its functions include muscle control and coordination, sensory reception and integration, speech production, memory storage, and the elaboration of thought and emotion.

LUNGS
Two sponge-like structures that fill most of the chest cavity are the **lungs**. Their essential function is to provide oxygen from inhaled air to the bloodstream and to exhale carbon dioxide.

LIVER
The **liver** lies largely on the right side of the abdominal cavity beneath the diaphragm. Its main functions are to process the contents of the blood and to break down fats, produce urea, filter harmful substances, and maintain a proper level of glucose in the blood.

BLADDER
The **bladder** is a fibromuscular organ located in the pelvic cavity. It stretches to store urine and contracts to release urine.

KIDNEYS
The **kidneys** are two bean-shaped organs located at the back of the abdominal cavity, one on each side of the spinal column. Their function is to maintain the body's chemical balance by excreting waste products and excess fluid in the form of urine.

HEART
The **heart** is a hollow, muscular organ that pumps blood through the blood vessels by repeated, rhythmic contractions.

STOMACH
The **stomach** is a muscular, elastic, pear-shaped bag, lying crosswise in the abdominal cavity beneath the diaphragm. Its main purpose is digestion of food through production of gastric juices.

INTESTINES
The intestines are located between the stomach and the anus and divided into two major sections: the **small intestine** and the **large intestine**. The function of the small intestine is to absorb most ingested food. The large intestine is responsible for absorption of water and excretion of solid waste material.

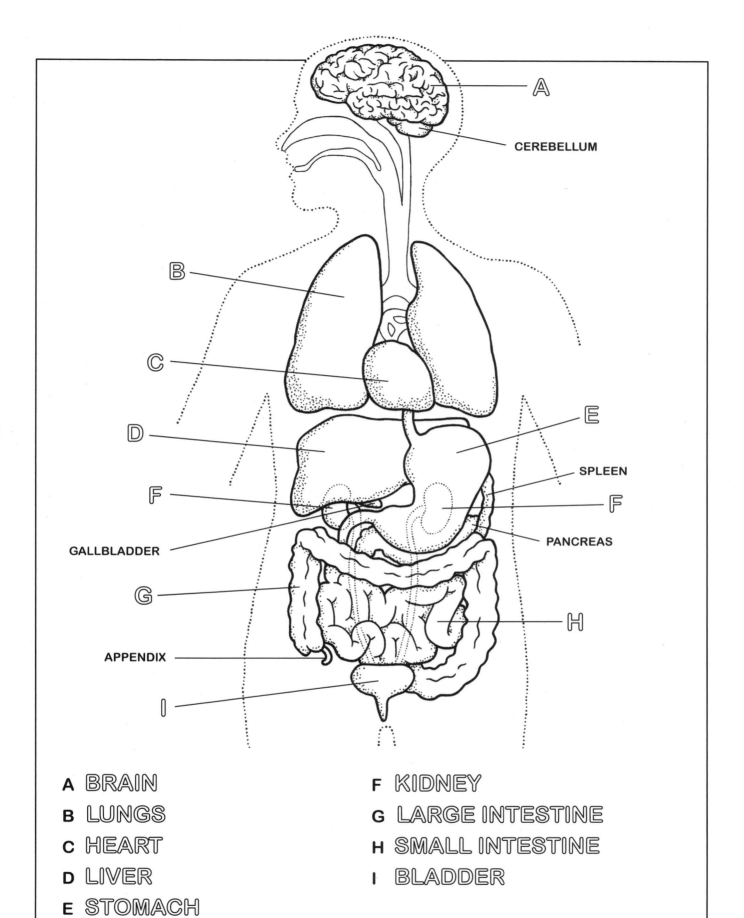

CEREBELLUM

SPLEEN

PANCREAS

GALLBLADDER

APPENDIX

A BRAIN
B LUNGS
C HEART
D LIVER
E STOMACH

F KIDNEY
G LARGE INTESTINE
H SMALL INTESTINE
I BLADDER

SECTION TWO:
THE TISSUES

The entire human body—consisting of trillions of cells, hundreds of organs, and many systems—demonstrates only four basic tissues. The tissues are the fabric of which the body is woven, and cells are its thread.

The classification of tissues is based on both morphology (shape) and function. Within each classification, there are several distinguishable types. It is helpful to understand these tissue types so that you can visualize macroscopic structures microscopically, and can then appreciate how the structure of an organ enhances or makes clear its function.

Generally, tissues are defined as collections of like cells, often attached to one another by fibers or intercellular "cement" and usually having a common function. The four tissues are usually combined in a way that an organ is formed. Consider the stomach: It is a cavity with walls. These walls consist of the four basic tissues, all working together to perform a common function, such as the initial phase of protein digestion.

The four tissues, such as in the stomach, are as follows:

• The lining of a cavity, consisting of a single layer of cells that secrete hydrochloric acid, mucous, and an enzyme—all of which aid in the chemical degradation of protein foodstuffs. Cells that line a cavity or any surface constitute epithelial tissue (epithelium). There are several varieties of epithelium.

• The material that supports the epithelium and gives it a degree of elasticity is appropriately termed connective tissue. It also supports local blood vessels and nerves, as well as glands and certain sensory receptors. Connective tissue comes in a variety of textures, densities, and compositions.

• The contractile matter that gives the stomach motility and provides for mechanical digestion is muscle tissue. Throughout the body, it comes in only three assortments.

• The excitable tissue that provides the stimulus for muscular contraction of the stomach is nervous tissue. Operating within complex interconnected networks, nervous tissue provides us with the ability to be not only sensitive to our environment but responsive to it as well.

In this section you will learn about the four basic tissues.

THE CELL

The **cell** is the fundamental unit of biological life—the common denominator of all living things. Any structure in the body more complex (e.g., bigger) than a cell is simply a *collection* of cells. Together, cells form tissues, tissues form organs, organs form organ systems, and organ systems combine to form an organism.

A cell is the anatomical sum of its parts. The parts involved in some basic task necessary for cell operation are called **organelles**. The parts that are products of cellular metabolism are called **inclusions**. Organelles and inclusions are made up of **organic** chemical compounds. The following classes of organic compounds are found in the cells: nucleic acids, proteins, lipids (fats), carbohydrates, and conjugated compounds. **Inorganic** compounds found in cells are generally salts, such as sodium chloride, potassium chloride, and calcium nitrate.

The cell is "ruled" by the **nucleus**, which is surrounded by a porous membrane. It is the administrative center in which the metabolic and synthetic activities of the cell are organized, initiated, and directed. The "blueprint" of every cell is the **DNA** (deoxyribonucleic acid) and this determines the size, shape, and function of a cell.

The outer part of the structure is the **cell** or **plasma membrane**. Each cell takes in substances with a cellular process called **endocytosis** through its membrane to form a **vacuole**. **Exocytosis** is a process by which a cell transports secreted material through the **cytoplasm** (the fluid that fills the cells) to the cell/plasma membrane. The **nuclear membrane** encloses the **nucleus** of the cell, and also connects the inner membrane with the outer membrane.

The **nucleoplasm** is a type of **protoplasm**, and is enveloped by the nuclear membrane. The nucleoplasm includes the **chromosomes** and **nucleolus**. The nucleolus, a round body located inside the nucleus, makes **ribosomal** subunits from proteins and ribosomal RNA, also known as rRNA. These rounded bodies, known as **ribosomes**, are derived from RNA of the nucleolus and are distributed throughout the **cytoplasm**.

Within the cytoplasm there is a network of tubules known as the **endoplasmic reticulum**, believed to serve as a mechanism for the movement of materials through the cell. They may be of a **smooth** or **rough** nature. The **Golgi Complex** (when the endoplasmic reticulum is near a complex of tubules and vesicles) is believed to receive newly synthesized protein from the endoplasmic reticulum and concentrate, package, store, and release it for final discharge in membrane-lined containers called **vesicles** or **vacuoles**. The secretory material in vesicles is moved through the cytoplasm to the free (unattached) border of the cell where, between fingerlike extensions (**microvilli** and **microfilament**) of the plasma membrane, it is released. These microvilli and microfilaments serve to increase the absorptive surface of the cell.

Foreign matter that crosses the cell membrane is met by **lysosomes**, which contain digestive materials and aid the cell in removing the unwanted particles.

Centrioles are shaped like small barrels. They are found in the cytoplasm in pairs. Their main function is to help with cell division, or **mitosis**. The centrioles help in the formation of the spindle fibers that separate the chromosomes during mitosis.

The energy for all cell activity comes from the structures called **mitochondria** (mitochondrion, sing.) which are found close to the site of any energy-consuming activity. Mitochondria are complex in organization and the site of enzymatic activity. They are the powerhouse of the cell and are responsible for cellular respiration and production. The mitochondria act like a digestive system that takes in nutrients, breaks them down, and creates energy-rich molecules for the cell.

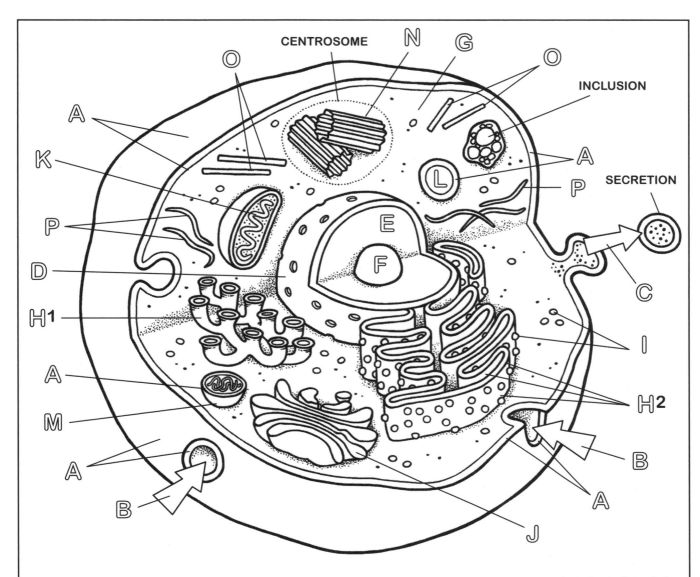

CENTROSOME N G O INCLUSION

O A

K A

P SECRETION

D

H1 C

A I

M H2

A B

B A

J

A CELL MEMBRANE

B ENDOCYTOSIS

C EXOCYTOSIS

D NUCLEAR MEMBRANE

E NUCLEOPLASM

F NUCLEOLUS

G CYPTOPLASM

H1 SMOOTH ENDOPLASMIC
RETICULUM

H2 ROUGH ENDOPLASMIC
RETICULUM

I RIBOSOME

J GOLGI COMPLEX

K MITOCHONDRION

L VACUOLE

M LYSOSOME

N CENTRIOLE

O MICROTUBULE

P MICROFILAMENT

CELL DIVISION & EPITHELIAL TISSUE

CELL DIVISION

One of the fundamental activities of all living organisms is reproduction. Reproduction is required for cellular replacement and tissue repair, and it is necessary for the growth of an organism and essential for the continuity of life. In humans, this reproductive process is started by the duplication of DNA, followed by a partial duplication and then exact division of cellular contents. This kind of cell division is called **mitosis**. In observing the stages of mitosis, one is primarily concerned with the changes taking place in the **nucleus**. The changes in the cytoplasm are apparent only in the final phase (**cytokinesis**) upon cleavage and formation of the two **daughter cells**. The phases of cell division are:

Interphase: Chromosomes are not visible but are dispersed as chromatin granules.

Prophase & prometaphase: 46 chromosomes condense and appear. Centrioles appear, each with a set of starlike rays (**asters**). Connecting the two centrioles are strands of fibers—the **spindle**. The centrioles move to opposite poles of the cell, stretching the spindle between them. The nucleolus disappears and the nuclear membrane disintegrates.

Metaphase: The 46 chromosomes line up along the middle of the cell in close relation to the spindle. Each chromosome is seen to split longitudinally, and each resultant half is called a **chromatid**, of which there are 92. The two halves of each chromosome attach to one another at the **kinetochore**, a disc-shaped protein structure that spindle fibers attach to.

Anaphase: Each chromatid splits off from its sister chromatid and they head to opposite poles of the cell. Free chromatids now constitute new chromosomes, and there are 46 at each of the two poles of the cell. The cytoplasm begins to cleave in the midline.

Telophase: A new nuclear membrane begins to appear around the chromosomes in each daughter cell. Chromosomes begin to disperse into chromatin. The two daughter cells mature into their final form, and the spindle fibers disappear. The two new daughter cells enter the interphase condition.

A INTERPHASE
B PROPHASE
C PROMETAPHASE
D METAPHASE
E ANAPHASE
F TELOPHASE
G CYTOKINESIS

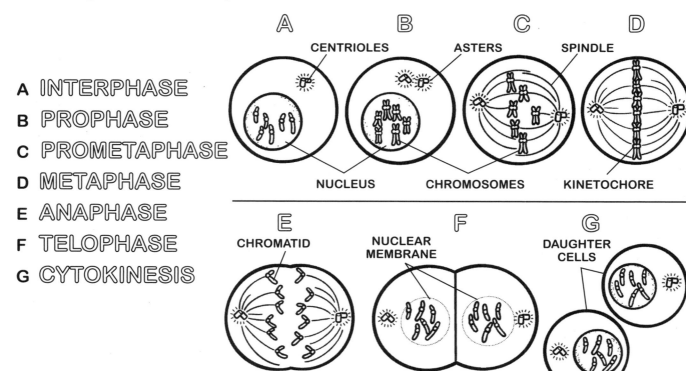

16

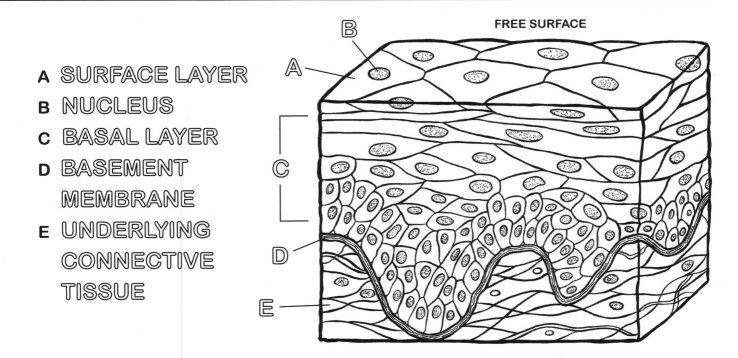

A SURFACE LAYER
B NUCLEUS
C BASAL LAYER
D BASEMENT MEMBRANE
E UNDERLYING CONNECTIVE TISSUE

FREE SURFACE

EPITHELIUM

Continuously growing or regenerating largely as a result of cell division, **epithelium** is one of the four basic types of tissue. It is a continuous, protective layer of densely packed cells, functioning as a barrier between the body and its environment. Epithelium is the **free** (unattached) surface of both closed and open cavities of the body, including all tubes, ducts, and vessels. In fact, there is an unbroken epithelial layer through your body. It extends without interruption in this sequence:

skin → mouth → throat → esophagus → stomach → intestines → rectum → anal canal → anus → external skin

In a way, you're like a human doughnut! There is a continuity of epithelium within and to all parts of your body by virtue of the millions of interconnecting blood vessels. The kinds of epithelium found in all these areas will differ according to functional requirements.

Epithelium that lines the interior of the blood vessels and the heart is called **endothelium**, and the epithelium making up the cellular lining of serous membranes is called **mesothelium**.

Epithelial tissue is generally characterized by the following:

• Cells are situated next to one another.
• There is little intercellular material.
• A free surface.
• Cells supported by and bound to the **underlying connective tissue** by a **basement membrane**.
• Absence of blood vessels (**avascularity**).
• Nutrition received by diffusion.

EPITHELIAL TISSUE
EPITHELIUM: CLASSIFICATION & FUNCTION

Epithelial tissue is characterized according to the shape of its cells and number of layers.

Squamous cells are thin, flat cells that look like fish scales and are found in the tissue that forms the surface of the skin, the lining of the hollow organs of the body, and the lining of the respiratory and digestive tracts. If the cells are in a single layer, the tissue is **simple squamous epithelium**. If the cells are in multiple layers, the tissue is **stratified squamous epithelium**.

Cuboidal epithelial cells are cube-shaped. In single layers, a tissue like this is known as **simple cuboidal epithelium**. In multiple layers, **stratified cuboidal epithelium**.

Cells shaped like rectangular columns are **columnar** cells. In a single-cell layer it is **simple columnar epithelium**, and in multiple layers, **stratified columnar epithelium**.

Aside from normal intracellular tasks, epithelial cells are capable of specialized activities that contribute to the function of a particular organ of which they are a part. In many cases, the function of the epithelial cell reflects the shape of the cell. Some of the functions carried out by epithelial tissue include:

• Protection against abrasion, ultraviolet light, etc.
• Filtration or diffusion.
• Absorption of material into a cell.
• Secretion of material to the outside of the cell.
• Sweeping material away by ciliary action.
• Flexibility, ability to withstand variable pressures and volumes.
• Reception of sensory stimuli, such as touch and pressure.

Consider the function of protection. What surfaces of the body require protection against abrasion, "weathering," wear-and-tear forces, and the like? Certainly the skin, for starters. And what is the epithelium lining the skin? **Stratified squamous epithelium**. Other surfaces requiring the services of this kind of epithelium include the mouth, anal canal, and vagina.

Simple squamous epithelium, on the other hand, cannot handle any trauma at all, and lines surfaces through which substances of molecular dimensions shuttle, as in filtration. It is found in such places as blood vessels and lymph vessels (endothelium), the filtering capsules and certain tubules of the kidney, and lining of closed cavities (mesothelium). The thinner the lining, the more improbable it is that protection of a wall or organ is one of its functions.

Cuboidal and **columnar epithelial tissue** is often single-layered and involved in the secretion and absorption of substances, as in the lining of the gastrointestinal tract, glands, and other places.

The respiratory structures—the trachea and related tubes, nasal cavity and such—require a **multifunctional epithelium** such as **pseudostratified columnar epithelium**, a single layer of cells that secretes mucous to trap foreign particles and has motile **cilia** (tiny hair-like structures) on its free surface to sweep the particles away to an orifice.

Certain parts of the kidney, the **ureters** (kidney-to-bladder tubes) and the urinary bladder, are subject to stretching due to variations in the volume of urine. The epithelium best adapted for this kind of stress is **transitional epithelium**.

A¹ SIMPLE SQUAMOUS BLOOD VESSELS, AIR SACS OF LUNGS

A² STRATIFIED SQUAMOUS
EPIDERMIS, VAGINA

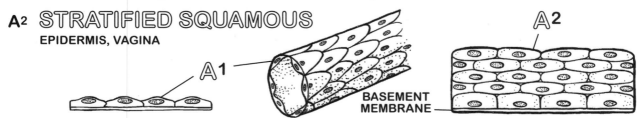

A¹

A²

BASEMENT MEMBRANE

B¹ SIMPLE CUBOIDAL
DUCTS AND SECRETORY PORTIONS OF SMALL GLANDS

B² STRATIFIED CUBOIDAL
SWEAT, SALIVA, MAMMARY GLANDS

BASEMENT MEMBRANE

B¹

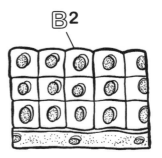

B²

C¹ SIMPLE COLUMNAR DIGESTIVE TRACT, BLADDER

C² STRATIFIED COLUMNAR MALE URETHRA, SOME GLANDS

C³ PSEUDOSTRATIFIED COLUMNAR BRONCHI, UTERINE TUBES, UTERUS

BASEMENT MEMBRANE

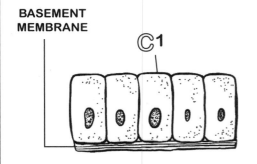

C¹

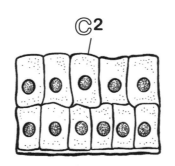

C²

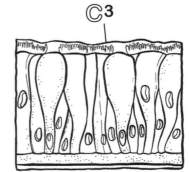

C³

D¹ TRANSITIONAL (STRETCHED)

D² TRANSITIONAL (NOT STRETCHED)
BLADDER, URETHRA, URETERS

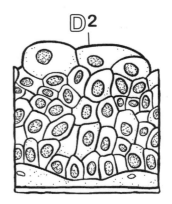

D²

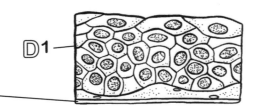

D¹

BASEMENT MEMBRANE

EPITHELIAL TISSUE
GLANDS

Epithelial cells and tissue whose major function is secretion are called **glands**. Glands may be composed of one cell (**unicellular**) or many cells (**multicellular**). They may secrete their product via a **duct** or **ducts** onto a surface, like **exocrine glands**, or directly into capillaries such as **endocrine glands**. Multicellular exocrine and endocrine glands develop from in-dippings or **invaginations**, an action or process to form a pouch or a cavity of surface epithelium. However, while the exocrine glands retain their connection with surface epithelium, the endocrine glands lose theirs.

Single cells specialized for secretion are **unicellular exocrine cells**, the most common of which are **goblet cells** in the epithelium of the digestive tube and the trachea. Scattered among the cells of simple epithelial tissues, they secrete their products directly on the free surface of open body cavities.

Multicellular exocrine glands make up the majority of glands found throughout the skin and mucous membranes. As you can see in the illustration, they are classified according to the number of ducts—**simple, branched,** or **compound**—and the shape of the secretory units; **alveolar, tubular,** or **coiled**.

Endocrine glands are ductless and secrete hormones directly into the blood. The pituitary, thyroid, adrenal, and other endocrine glands will be further discussed in the context of the region of the body where they are found.

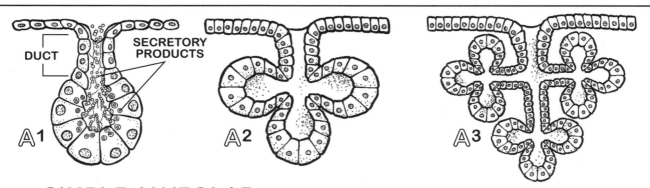

A¹ SIMPLE ALVEOLAR A STAGE OF DEVELOPMENT FOR SIMPLE BRANCHED GLANDS

A² SIMPLE BRANCHED ALVEOLAR SEBACEOUS GLANDS

A³ COMPOUND ALVEOLAR MAMMARY GLANDS

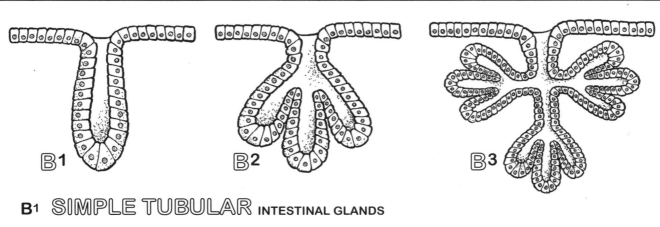

B¹ SIMPLE TUBULAR INTESTINAL GLANDS

B² SIMPLE BRANCHED TUBULAR GASTRIC AND SOME MUCOUS GLANDS

B³ COMPOUND TUBULAR MUCOUS GLANDS IN MOUTH, MALE REPRODUCTION

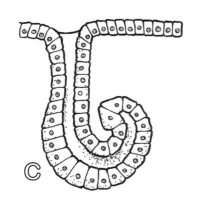

C SIMPLE COILED TUBULAR

MEROCRINE SWEAT GLANDS

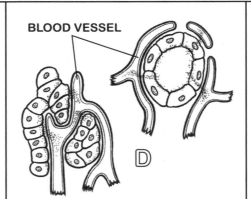

D ENDOCRINE GLANDS

PITUITARY, THYROID, ETC.

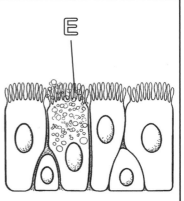

E GOBLET CELL

TRACHEA, ETC.

CONNECTIVE TISSUE:
CELLS & FIBERS

In the overall sense, connective tissue may be defined as a variety of cellular aggregations and their secretory products which generally function to bind, support, or link other structures in the body. Put more simply, connective tissue *connects*. All organs are bound in connective tissue to one degree or another—the body is supported by this tissue. It occupies more space in the body than any other tissue.

While epithelial tissue is composed of closely packed cells with little or no extracellular space between them, connective tissue cells are dispersed within an **extracellular matrix**, which helps bind the cells together and regulates various cellular functions. The matrix can have a viscous or jelly-like consistency, made up of primarily protein-sugar complexes (**mucopolysaccharides**), or it can be mineralized and solid.

The classification of connective tissue is based on the relative density and character of this extracellular matrix. Connective tissue that consists of cells and fibers of variable ratios in a fluid matrix is called **connective tissue proper**. Connective tissue that consists of cells and fibers embedded in a solid substance is called **supporting tissue** (cartilage and bone). **Blood** and **lymph** are connective tissues in which the extracellular material is liquid and which contain a variety of cells and compounds. We will explore these categories of connective tissue on the following pages.

CELLS IN CONNECTIVE TISSUE
A variety of cells may be found in connective tissue, with the two most common depicted here.

Fibroblast are the source of all connective tissue fibers and are found in fibrous connective tissues. Their primary function is to maintain connective tissue's structural integrity by continuously secreting precursors of the **extracellular matrix**: the substance within the extracellular space and a variety of fibers. Fibroblast are especially important in secreting collagenous fibers at an inflammation site, to wall off the inflammation.

Macrophages are cells that "eat up" unwanted viruses, bacteria, injured cells, or foreign material in a process called **phagocytosis**. They engulf the unwanted particle, break it down with enzymes stored in their **lysosomes**, and dispose of any leftover waste. They are often difficult to identify unless filled with ingested material.

Other cells of the connective tissues include certain white blood cells, fat cells, antibody-producing cells (plasma cells), and heparin- and histamine-secreting cells (mast cells).

FIBERS IN CONNECTIVE TISSUE
The fibers of connective tissues are of three assortments.

Collagenous (white): Present in all connective tissue including bone, these fibers are made of the protein collagen and have the tensile strength of steel.

Elastic: Present in most connective tissue to one degree or another, these impart a yellow color if present in large concentrations. Elastic fibers make the principally collagenous tissues elastic and flexible.

Reticular: These fibers are small and delicate in proportions, forming networks to support cells.

FIBROBLAST SECRETIONS

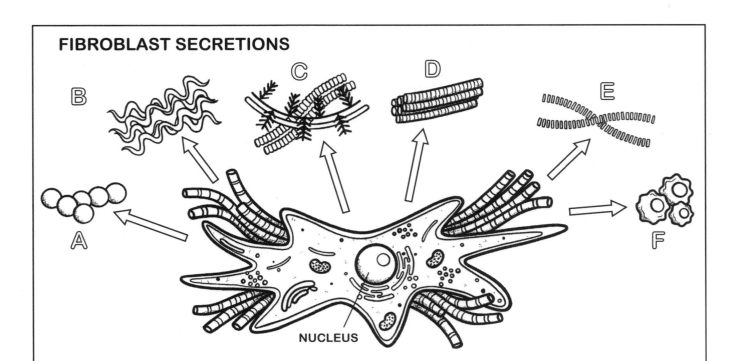

NUCLEUS

A HYALURONIC ACID
B ENZYMES
C EXRACELLULAR MATRIX

D COLLAGEN FIBRILS
E ELASTIN
F GROWTH FACTORS

PHAGOCYTOSIS IN SIX STEPS

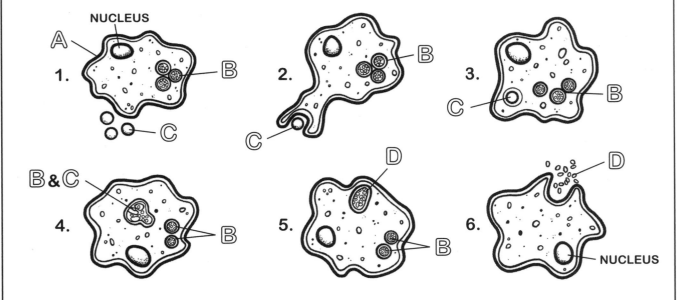

NUCLEUS

1. 2. 3.

4. 5. 6.

NUCLEUS

A MACROPHAGE
B LYSOSOMES

C VIRUSES/BACTERIA
D WASTE MATERIAL

CONNECTIVE TISSUE
CONNECTIVE TISSUE PROPER

Connective tissue proper comes in several varieties.

Loose is characterized by having many cells and fewer fibers. This most common of all connective tissue fills any and all unoccupied spaces in the body, binds all skeletal muscle, and makes up the connective tissue layers in mucous membranes, as well as the subcutaneous tissue "holding down" the skin.

Dense characteristically has many fibers. The chief functional feature here is strength. The fibers may be arranged in coarse, interwoven bundles, as in **dense irregular**, or arranged neatly in parallel, as in **dense regular**. Dense irregular is significantly stronger than loose and is found encapsulating many organs of the body. As periosteum it covers bone; as perichondrium it covers cartilage. Dense regular is particularly suited for tissues that experience high tensile (pulling) forces, such as ligaments (which connect bone to bone) and tendons (which connect muscle to bone or muscle to muscle).

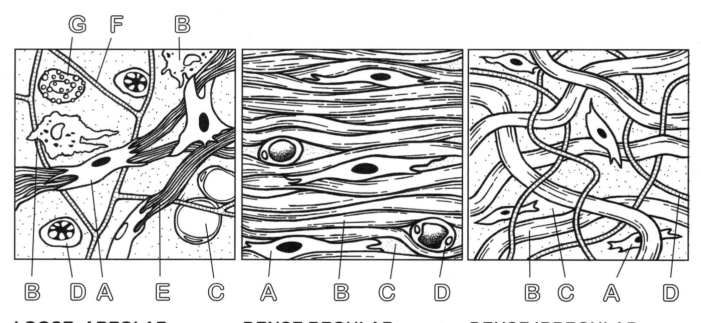

LOOSE, AREOLAR

A FIBROBLAST

B MACROPHAGE

C FAT CELL

D PLASMA CELL

E COLLAGEN FIBERS

F ELASTIC FIBERS

G MAST CELL

DENSE REGULAR

A FIBROBLAST

B COLLAGEN FIBERS

C MATRIX

D CAPILLARY

DENSE IRREGULAR

A FIBROBLAST

B COLLAGEN FIBERS

C MATRIX

D ELASTIC FIBERS

Elastic is found in the walls of large arteries and in certain ligaments. Elastic offers a degree of resiliency not found in other tissues.

Adipose is composed of large spherical cells whose cytoplasm is filled with fat. Thus it is called fatty tissue. The cells are often pressed together and the nuclei of the cells are pushed off to one side. Fatty tissue is active metabolically; its ability to synthesize fat from carbohydrates is influenced by hormones. Fatty tissue is fairly rich in its supply of nerves and blood vessels.

Reticular consists of delicate fibers and cells that form a supportive network for free and fixed cells composing such structures as lymphatic organs, bone marrow, and liver.

Embryonic is the connective tissue found in the embryo (not pictured).

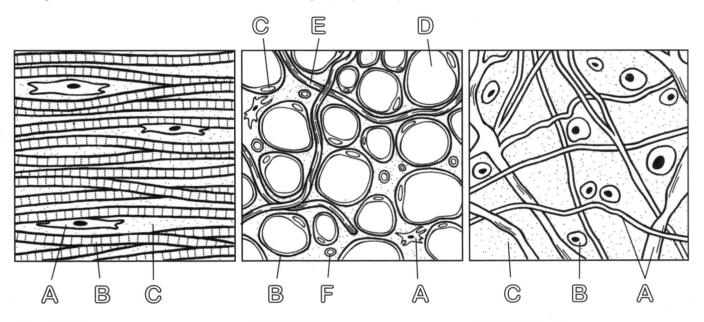

ELASTIC

A FIBROBLAST
B ELASTIC FIBERS
C MATRIX

ADIPOSE

A FIBROBLAST
B FAT CELL
C NUCLEUS
D LIPID DROPLET
E SUPPORTING FIBERS
F CAPILLARY

RETICULAR

A RETICULAR FIBERS
B CELLS OF THE ORGAN
C MATRIX

CONNECTIVE TISSUE
SUPPORTING TISSUE: CARTILAGE

Supporting connective tissue that occurs either as **bone** or **cartilage** is characterized by a dense, hard matrix, which gives it the ability to support weight. Cartilage is less hard than bone because it lacks the minerals (calcium crystals) of bone—a little like a cement mix without the cement! Cartilage consists of **chondrocyte** cells enclosed in a small cavity (**lacuna**) all within a matrix that is supported by collagenous fibers, much as concrete is supported by steel rods.

Cartilage grows faster than bone but slower than connective tissue proper. Cartilage, while hard, is somewhat flexible and is responsible for supporting the embryo during its development. Cartilage makes up the skeleton of the embryo and is progressively replaced by bone beginning in the fetus and ending in the young adult, in whom the remaining skeletal cartilage occupies only the ends of long bones (articular surfaces) in the skeleton.

There are three types of cartilage:

Hyaline cartilage has a glassy appearance with a blue tint. When cut, it chips like ice. It is found at the end of bone (articular surfaces); in the nose, the trachea, and parts of the larynx; and it composes the cartilaginous portions of the ribs. Try to feel hyaline cartilage on yourself: Bend your nose back and forth—the supporting framework of this proboscis is hyaline cartilage. Now run your fingers over the cartilages of the "Adam's apple" and other parts of the larynx and the trachea below that. Compare the density of these cartilages with the bones of your hand.

The transition zone between hyaline cartilage and its loose binding is a dense fibrous tissue called **perichondrium**.

Elastic cartilage is a hyaline cartilage infiltrated with elastic fibers, thus allowing the tissue to become flexible and resilient. Elastic cartilage can be explored on yourself: Twist and bend your external ear.

Fibrous cartilage or **fibrocartilage** is really dense regular with cartilage cells (chondrocytes) interspersed. Frequently it seems that fibrocartilage is a transition tissue between hyaline cartilage and the surrounding dense connective tissue (regular or irregular). Fibrocartilage is found between adjacent vertebral bodies of the spinal column, between the pubic bones of the hip, and in most transitional tissues between joint capsules, ligaments, tendons, and articular cartilage.

Cartilage lacks blood vessels and nerve fibers. The chondroblasts and chondrocytes synthesize and secrete the fibers and matrix of cartilage.

Bone is the most dense and hardest of all the connective tissues, and in fact, of all tissues of the body except teeth.

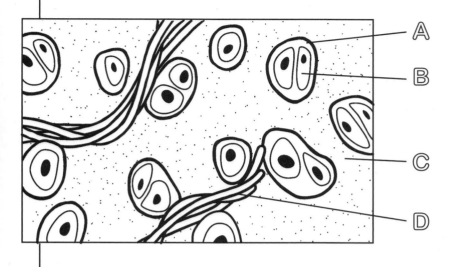

HYALINE CARTILAGE

A LACUNA
B CHONDROCYTE
C MATRIX
D COLLAGEN FIBERS

ELASTIC CARTILAGE

A LACUNA
B CHONDROCYTE
C MATRIX
D ELASTIC FIBERS

FIBROCARTILAGE

A LACUNA
B CHONDROCYTE
C MATRIX
D COLLAGEN FIBERS

CONNECTIVE TISSUE
SUPPORTING TISSUE: BONE

Bone carries out a number of important structural and physiological activities:

• It serves as an internal skeleton for support of the body.
• It is the site of attachment for muscles, tendons, and ligaments.
• It offers protection of cranial, thoracic, and pelvic viscera.
• It is a center of blood-forming (hemopoietic) activity.
• It is a source of calcium needed by the body.

Bone is composed of:

• **Osteoblasts**: Developing bone cells that form new bone.
• **Osteocytes**: Bone cells formed when an osteoblast becomes embedded in the matrix it has secreted.
• **Osteoclasts**: Bone-destroying cells responsible for aged bone resorption. Osteoclasts and osteoblasts are linked in the lifelong process of bone remodeling.
• **Organic ground substance**: Collagenous fibers, mucopolysaccharides.
• **Inorganic (mineral) substance**: Calcium and phosphate complexes.

Osteogenic cells are **stem cells**—cells with the ability to develop into different types of specialized cells. They are the only bone cells that divide. They differentiate into osteoblasts and **chondroblasts**, which will form chondrocytes and secrete extracellular matrix to maintain cartilage.

Based on structural differences, there are two types of bone:

• **Compact**, or dense.
• **Cancellous**, or spongy.

These two types are essentially the same in constitution; the difference lies in the organization of the tissue.

In compact bone, the tissue is densely packed into cylindrical structures called **osteons** or **haversian systems**. Within their concentric rings of matrix, osteons have a central channel called the **haversian canal**, surrounding blood vessels, and nerve cells that facilitate communication in bone cells.

In cancellous bone, the tissue is arranged into thin **trabeculae**—tiny supportive struts with spaces between and among them. Haversian systems are absent in cancellous bone.

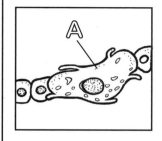

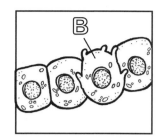

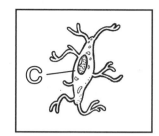

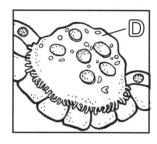

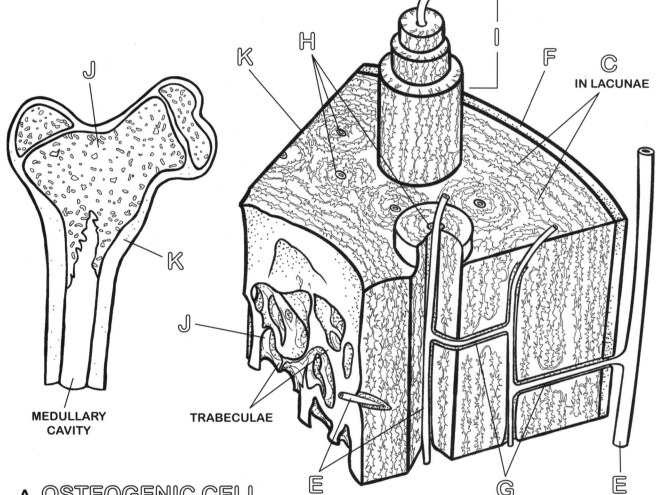

IN LACUNAE

MEDULLARY
CAVITY

TRABECULAE

A OSTEOGENIC CELL

B OSTEOBLAST

C OSTEOCYTE

D OSTEOCLAST

E BLOOD VESSEL

F PERIOSTEUM

G PERFORATING CANAL

H HAVERSIAN CANALS

I OSTEON

J CANCELLOUS BONE

K COMPACT BONE

CONNECTIVE TISSUE
BONE DEVELOPMENT: PHASE ONE

The functional relationships of the cells and other structures associated with bone can best be appreciated in the development of bone—a process taking twenty to twenty-five years in a human being. The process of bone development (referred to as **osteogenesis**) and formation (**ossification**) begins during embryonic growth in the first few weeks after conception. Bone may develop directly from **mesenchyme**, a type of connective tissue found mostly during embryonic development and comprised of loose cells embedded in the extracellular matrix. This interaction is **intramembranous ossification**. Formation of bone may also develop by replacing the cartilage described earlier (termed **endochondral ossification**). All the bones of the body develop by the latter process with the exception of the flat bones of the skull and the clavicle.

PROCESS OF INTRAMEMBRANOUS OSSIFICATION

Osteoblasts (developing bone cells) differentiate from **mesenchymal cells**. Osteoblasts form clusters and start secreting unmineralized bone—**osteoid** tissue—which will mature as bone. Some osteoblasts have become trapped in their own secretions. These cells are now **osteocytes** and reside in spaces called **lacunae**. The surrounding mesenchyme will differentiate into more osteoblasts or **periosteum**.

In the adjacent illustration, note that the periosteum has formed on both sides of developing flat bone. The major tissue here is now bone—the osteoid tissue is becoming mineralized. Note the osteocyte in their lacunae. Large spaces at the sides and in the center are marrow cavities containing blood vessels—arterioles and veins—osteoblasts, and **osteoclasts**.

Intramembranous bone, even if fully developed, is reabsorbed by osteoclasts and replaced by osteoblasts throughout life.

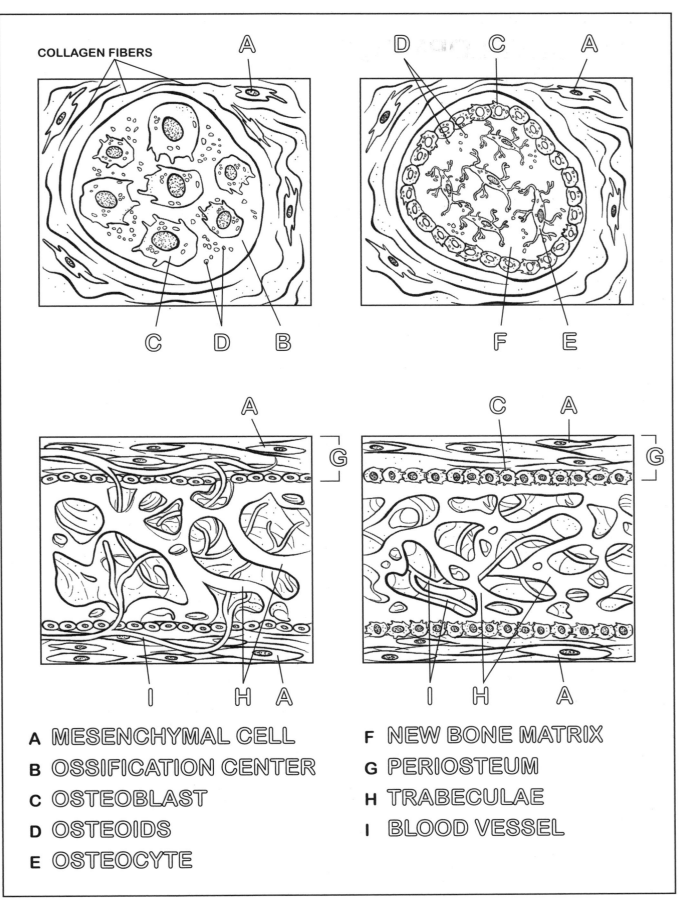

COLLAGEN FIBERS

A MESENCHYMAL CELL
B OSSIFICATION CENTER
C OSTEOBLAST
D OSTEOIDS
E OSTEOCYTE

F NEW BONE MATRIX
G PERIOSTEUM
H TRABECULAE
I BLOOD VESSEL

CONNECTIVE TISSUE
BONE DEVELOPMENT: PHASE TWO

Formation of bone may also develop by **endochondral ossification**, a process by which growing hyaline cartilage is systematically replaced by bony tissue to form the growing skeleton. Most of the bones of the body develop by this process.

PROCESS OF ENDOCHONDRAL OSSIFICATION

In the early stages of embryonic development, skeletons of cartilage are formed from mesenchyme, and these cartilaginous models have the basic shape of the adult bones that later develop.

Bone cells arise in a separate line of development from mesenchyme cells; therefore in endochondral ossification, bone cells must and will replace the cartilage cells.

The **perichondrium** is a layer of fibrous connective tissue that covers the cartilage and is the source of chondroblasts (transformed from mesenchymal cells) which add to the growing cartilage by **appositional growth**—the addition of bone tissue at the bone's surface.

The **chondrocyte cells** within the cartilage mature and enlarge. As they do, the matrix thins out and they lose much of their source of nutrition, particularly in the central zone of the shaft. The chondrocytes enlarge and then die. Thin spicules (trabeculae) of the matrix begin to absorb calcium salts and become a **calcified matrix**. Calcification of cartilage is, in a sense, a pathological process. In this case, it is necessary so that bone may replace it.

While the spicules within the central zone of cartilage calcify, the perichondrium is invaded by capillaries. This event influences the mesenchymal cells of the inner layer of perichondrium to differentiate into osteoblasts (as well as chondroblasts). The perichondrium now becomes known as **periosteum** and a layer of bone is secreted along the sides of the central zone of the shaft. This layer is the **bone collar**.

The bone collar supports the shaft of the cartilage while the central zone, now filling with calcium crystals, deteriorates and disintegrates. Capillaries of the periosteum grow into the central zone. These are **periosteal buds** and carry with them osteoblasts and undifferentiated mesenchymal cells.

The central zone of calcified spicules (trabeculae), periosteal bud, osteoblasts, and embryonic bone constitute a **diaphyseal (shaft) center of ossification**. The osteoblasts line up around trabeculae of calcified cartilage and secrete osteoid tissue on them. The calcified cartilage disintegrates and is absorbed into the capillaries. Osteoid tissue remains and is subsequently mineralized to become bone.

Please note the progressive ossification along the sides of the shaft is to be **compact bone**, and the center of the bone becomes **cancellous bone**. As the bone collar increases in length and thickness, the interior cancellous bone dissolves to a large extent, leaving the **medullary cavity** (*medulla* meaning innermost part) where red and/or yellow bone marrow is stored; hence, it is also known as the **marrow cavity**.

The process of chondrocyte hypertrophy and calcification of cartilage begins in the ends of the long bone as it occurred previously in the center. Capillary invasion brings in more periosteal buds and starts **new centers of ossification** in the rounded ends of a long bone (the **epiphysis**, plural **epiphyses**) and the process of ossification continues to spread out to just short of the outer limit of the cartilage—which will remain throughout life as the **articular cartilage**.

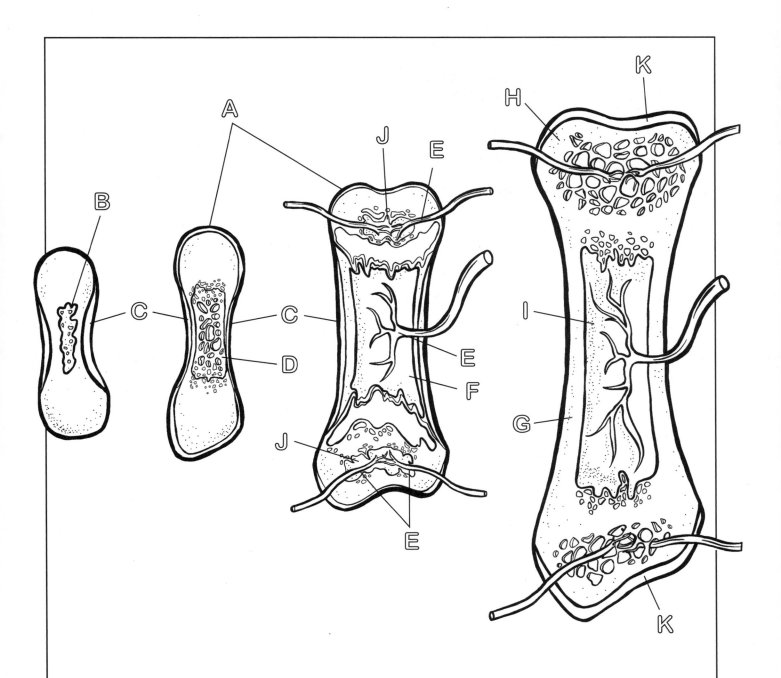

A PERICHONDRIUM

B CALCIFIED MATRIX

C PERIOSTEUM **OR**
BONE COLLAR

D DETERIORATING
CARTILAGE MATRIX

E PERIOSTEAL BUD

F CENTER OF OSSIFICATION

G COMPACT BONE

H CANCELLOUS BONE

I MEDULLARY CAVITY

J NEW CENTERS OF
OSSIFICATION

K ARTICULAR CARTILAGE

CONNECTIVE TISSUE
BONE GROWTH SUMMARY &
THE ANATOMY OF A BONE

To summarize bone development makes it easier to understand how human bones grow:

• Bones are preceded by cartilage models.

• The bone collar surrounds and supports the central shaft—the **diaphysis**—while the cartilage of the central zone is disintegrating.

• The periosteal bud brings into the central zone the cellular elements necessary for ossification.

• The key feature of the ossification process is replacement of calcified cartilage by bone.

• The formation of the epiphyseal centers of ossification (in the **epiphyses** of the bone) makes possible the mechanism for lengthening bone growth.

• The **epiphyseal plate** is a zone of cartilage between the epiphyseal and diaphyseal centers of ossification, which is in a losing race with those centers in terms of rate of **mitosis**—a method of cell division in which a cell divides and produces identical copies of itself. As a consequence, bone lengthens.

The epiphyseal plate separates epiphyseal bone from diaphyseal bone and in fact, because of its continued growth, pushes the epiphysis away from the diaphysis.

The epiphyseal plate, however, does not become larger or thicker. It becomes smaller and thinner! The rate of proliferation of chondrocytes in the epiphyseal plate is, in a sense, in a race with the rate of proliferation of developing bone on the diaphyseal side of the plate. The proliferation of cartilage in the plate makes the plate thicker. A proliferation of developing bone on the diaphyseal surface of the plate (at the expense of the epiphyseal plate cartilage) makes the plate thinner.

Even after growth of bone ceases, reabsorption and replacement of bone continues throughout life. These activities provide for possible change in the architecture of cancellous (spongy) bone to compensate for certain stresses, e.g., increased body weight, increased muscle pull, nutritional deficiencies, postural changes, and fractures. As a result of the reabsorption of the mineral elements of the bone, calcium is set free as ions for absorption by the blood—mineral levels of calcium in the blood are essential for life.

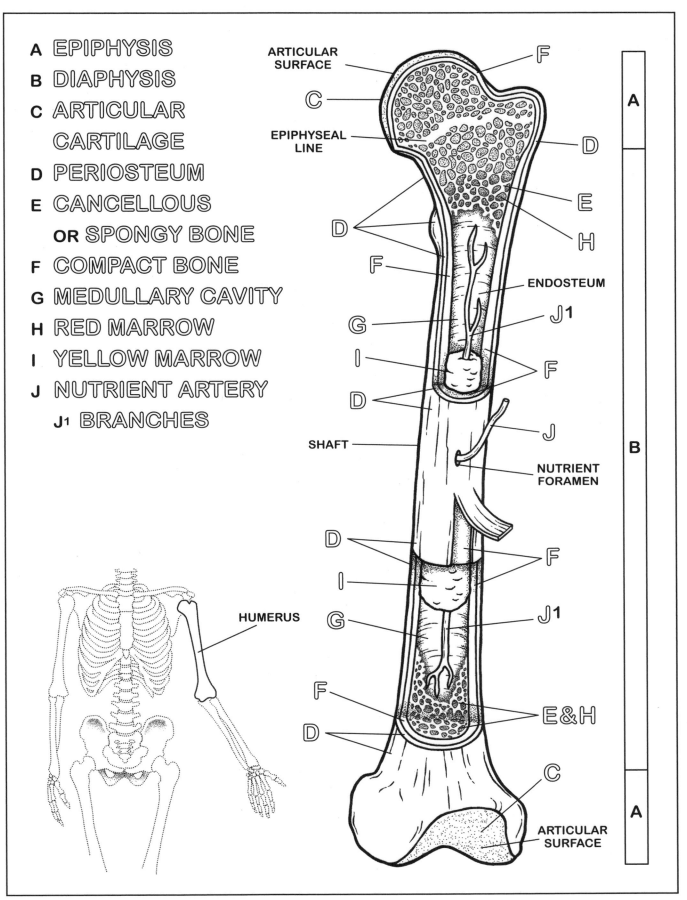

A EPIPHYSIS
B DIAPHYSIS
C ARTICULAR
   CARTILAGE
D PERIOSTEUM
E CANCELLOUS
   OR SPONGY BONE
F COMPACT BONE
G MEDULLARY CAVITY
H RED MARROW
I YELLOW MARROW
J NUTRIENT ARTERY
   J1 BRANCHES

ARTICULAR SURFACE
EPIPHYSEAL LINE
ENDOSTEUM
NUTRIENT FORAMEN
SHAFT
HUMERUS
ARTICULAR SURFACE

35

CONNECTIVE TISSUE
RED BLOOD CELLS (RBCs)

Blood is also connective tissue, consisting of fluid as well as a cellular phase.

The fluid phase of blood is called **plasma**, a type of extracellular fluid (ECF) that makes up about 5 percent of the body weight. Plasma contains an unknown but great number of ions and molecules—both organic and inorganic—which easily communicate with other fluid spaces of the body. Digested foodstuffs of molecular dimensions, drugs, hormones, and a variety of protein enzymes, antibodies, and other substances regularly pass into and out of the plasma. Infective agents also employ the plasma as a vehicle for spreading to other areas.

If plasma is allowed to stand in air, it will clot, and the fluid remaining after the clot is **serum**. Since the clot is the result of a complex series of reactions involving the plasma proteins, serum differs from plasma in having fewer such proteins as well as an absence of other clotting substances.

If blood is centrifuged, or is allowed to stand without clotting, the plasma separates for the cellular portion. The plasma phase is straw-colored. Plasma and serum isolated from blood samples taken from patients may reveal symptoms of a disease or destructive process underway upon being tested in the clinical laboratory by medical technologists.

The cellular phase of blood makes up about 45 percent of the total blood volume. The constituents of this phase are called formed elements and they consist of:

• **Erythrocytes**, or red blood cells (RBCs), 44 percent by volume of blood.
• **Leukocytes**, white blood cells (WBCs), 1 percent by volume of blood.
• **Platelets**, less than 1 percent by volume of blood.

All of these elements develop from stem cells in the bone marrow or lymphatic organs.

ERYTHROCYTES: RED BLOOD CELLS (RBCs)
Erythrocytes are the most commonly seen elements in blood; there are about 5 million per cubic millimeter! They are the remains of cells that extruded their nuclei in the latter phases of development, hence they are not called cells; *corpuscles* or *elements* would be more accurate.

An erythrocyte is a biconcave disc measuring about six to eight micrometers (**μm**) in diameter. In essence, each is a sac of **hemoglobin**, the protein that carries oxygen from the lungs to the body's tissues, and then returns carbon dioxide from the tissues back to the lungs. Hemoglobin is made up of four subunits, each of which contains **iron atoms**—vital in transporting oxygen and carbon dioxide to and from the tissues. The iron contained in hemoglobin is also responsible for the red color of blood. Hemoglobin is enclosed by a plasma **membrane**.

Thus the primary functions of the erythrocytes are: (1) to combine with oxygen inhaled into the lungs (hemoglobin + oxygen = oxyhemoglobin) and (2) to release that oxygen at the tissue level (by a process of diffusion). A reduction in quality or quantity of RBCs in the blood, for any reason, is termed **anemia**. An excess of RBCs in the blood constitutes **polycythemia**. The former is much more common than the latter.

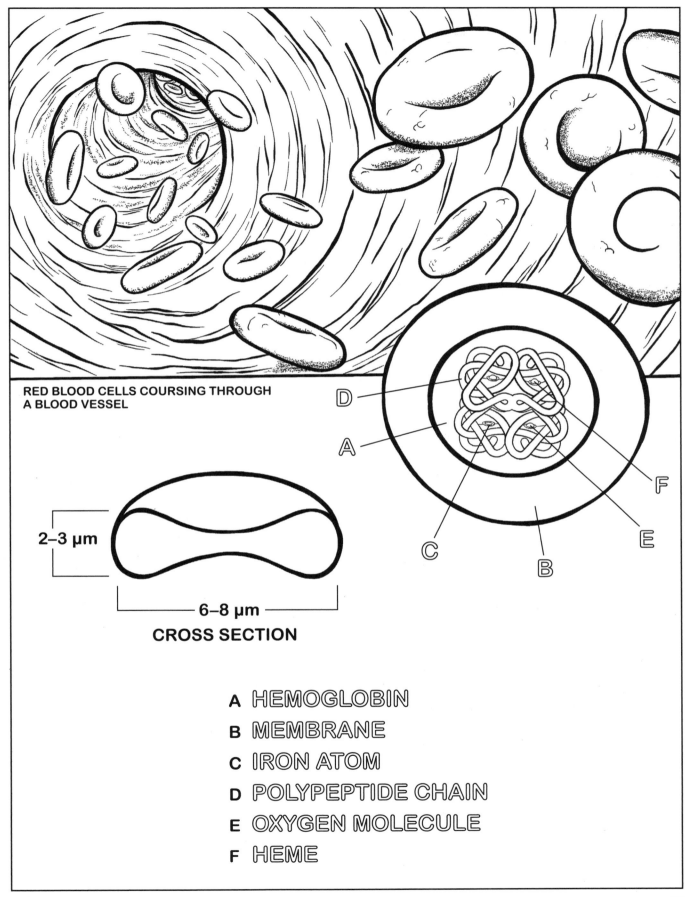

RED BLOOD CELLS COURSING THROUGH
A BLOOD VESSEL

2–3 μm

6–8 μm

CROSS SECTION

A HEMOGLOBIN

B MEMBRANE

C IRON ATOM

D POLYPEPTIDE CHAIN

E OXYGEN MOLECULE

F HEME

CONNECTIVE TISSUE
WHITE BLOOD CELLS (WBCs)

White blood cells are **leukocytes**, the truly cellular elements of the blood. They make up only about 1 percent of the total blood volume—but a vital 1 percent they are! They average about 8,000 cells per cubic millimeter of blood. A significantly decreased number of leukocytes constitutes **leukopenia** (-*penia*, poverty) while an increased number is **leukocytosis**. An increased number of a particular cell type due to a malignant process (**neoplasm**) is generally called **leukemia** (-*emia*, in the blood).

Leukocytes are generally larger than RBCs (10 to 16 μm in diameter) and are not at all concerned with oxygen transport. The approximate percentage of total leukocytes in the blood are:

- **Neutrophils** 65%
- **Eosinophils** 3%
- **Basophils** 1%
- **Lymphocytes** 30%
- **Monocytes** 1%

Neutrophils, eosinophils, and basophils are collectively termed **granular leukocytes** because there are obvious granules in their cytoplasm. The last two leukocytes are termed **agranular** (nongranular) leukocytes. Neutrophils are so called because they stain with a neutral (not acidic or basic) stain. They are highly mobile and can migrate out of circulation into the connective tissue spaces to help combat an infection by phagocytizing bacteria. They may also be involved in immune reactions to foreign substances where specific antibodies play a major role.

Eosinophils are named because their cytoplasmic granules stain eosin, an acid dye. Eosinophils destroy invading germs like viruses, bacteria, or parasites such as hookworms. They also have a role in the inflammatory response, especially if an allergy is involved.

Basophils are so called because their cytoplasmic granules stain with a basic dye. Their primary function is to help the body fight off inflammatory reactions in some acute and chronic allergic diseases such as asthma, dermatitis, and hay fever. They are known to contain significant **titers** (concentrations) of heparin, histamine, and serotonin.

Lymphocytes arise from lymphatic tissue (found in the spleen, lymph nodes, tonsils, and nodules of the mucous membranes). Lymphocytes are made from stem cells in the bone marrow. They are generally nongranular when seen under the microscope, but occasionally contain sky-blue granules. Lymphocytes come in large, medium, and small sizes, with only the latter two found in the blood. They are known to secrete antibodies that counteract the effect of **antigens**—foreign substances in the bloodstream. There are two main types of lymphocytes: **T cells** and **B cells**. T cells help the body kill cancer cells and control the immune response to foreign substances, and B cells produce antibody molecules that can latch on to and destroy invading viruses or bacteria.

Monocytes are the largest of the white blood cells and are believed to be phagocytic. Monocytes differ from lymphocytes in that they are larger in size and have a kidney-shaped nucleus.

Platelets are pinched-off cytoplasmic bodies of **megakaryocytes** (cells found in bone marrow) and play a major role in blood clotting.

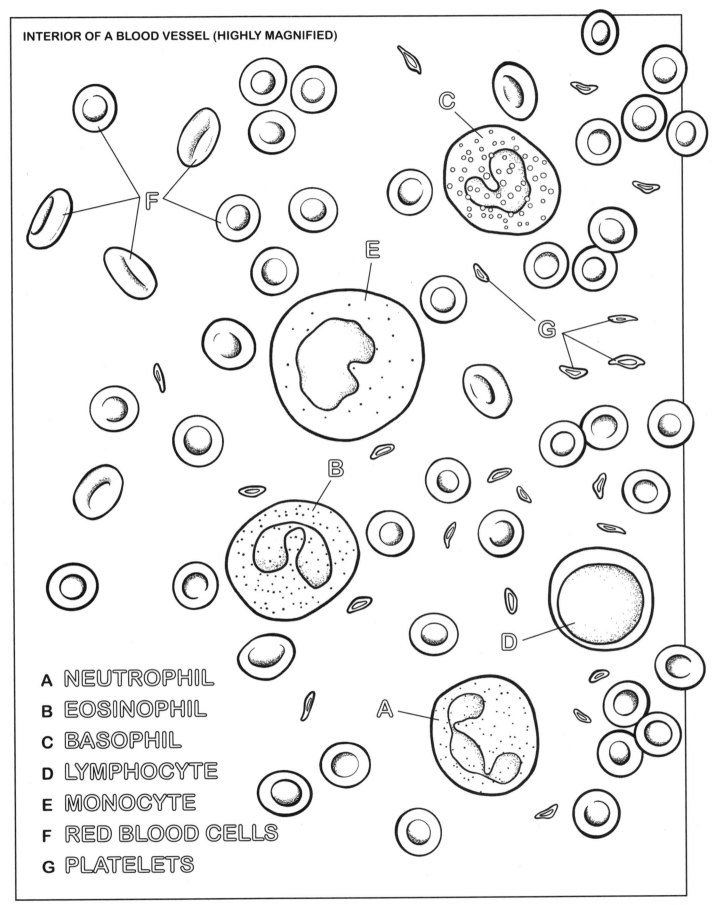

INTERIOR OF A BLOOD VESSEL (HIGHLY MAGNIFIED)

A NEUTROPHIL
B EOSINOPHIL
C BASOPHIL
D LYMPHOCYTE
E MONOCYTE
F RED BLOOD CELLS
G PLATELETS

CONNECTIVE TISSUE
LYMPH

Lymph is extracellular fluid (ECF), found in the lymphatic vessels of the body. It is a whitish, milky fluid and there are large fatty molecules and small proteins as well as lymphocytes in it.

Lymph is generated in the tissue fluids and diffuses into **lymph capillaries**, as seen here. Lymph then circulates through the body using a network of lymphatic vessels. Distributed throughout this network are **lymph nodes**, small balls of lymphatic tissue that work as filters for harmful substances. As the lymph circulates through the lymph nodes, lymphocytes are added that can help fight infection.

The lymphatic system helps rid the body of toxins, waste, and other unwanted materials. Ultimately, the lymph empties into one of two large veins draining into the heart.

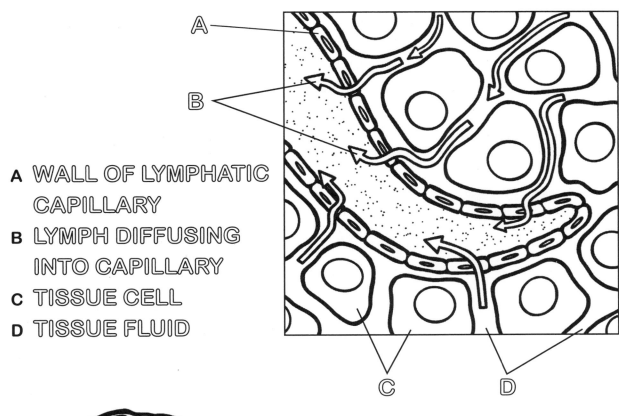

A WALL OF LYMPHATIC
   CAPILLARY
B LYMPH DIFFUSING
   INTO CAPILLARY
C TISSUE CELL
D TISSUE FLUID

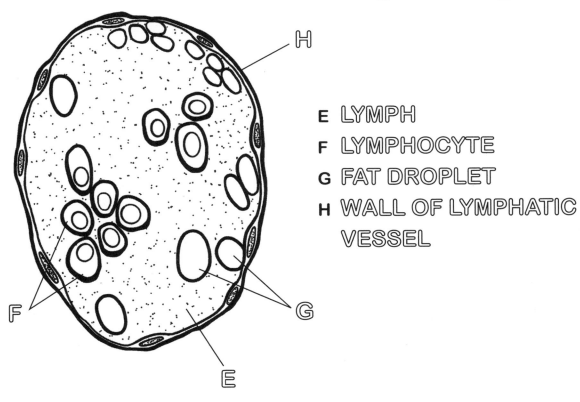

E LYMPH
F LYMPHOCYTE
G FAT DROPLET
H WALL OF LYMPHATIC
   VESSEL

CONNECTIVE TISSUE
MUSCLE TISSUE

Certainly if connective tissue takes up more of your body than any other tissue, **muscle tissue** must run a close second! For if any part of you moves—muscle moves it. Think about that for a moment. Several muscles are involved in just tapping your foot or scratching your head—not to mention running or walking. How about the twitches you get occasionally? Swallowing, breathing, digestion, defecation, urination? All these movement-based processes occur because muscles **contract**—they shorten or undergo an increase in tension.

Aside from obvious muscle movement, there are many muscular contractions you are rarely aware of: contractions of the heart (the heart is practically all muscle); contractions of muscle in blood vessels so your blood pressure will rise when it's necessary; most of the muscular contractions of the digestive tract; sustained or **tonic** contraction of postural muscles to keep you erect when sitting or standing.

By definition, muscle tissue is a contractile tissue composed of elongated cells, often attached to one another through the medium of connective tissue fibers. Muscle cells are frequently called **fibers** because of their external appearance; however they are "living fibers" and not at all "fibrous" in the connective tissue sense of the word.

Muscle tissues are classified on the basis of their microscopic appearance. Those muscle cells exhibiting cross striations throughout their full length are **striated muscle**. Two types of striated muscle may be recognized: Striated muscle associated with the bony skeleton of the body is known as **skeletal muscle**, which moves and stabilizes the various positions of the skeleton; and striated muscle making up the muscular component of the heart is **cardiac muscle**, which moves and maintains blood pressure. Muscle cells lacking cross striations are **smooth muscle**. Smooth muscle is a constituent of visceral walls, moving secretions from food, reproductive, and urinary tracts, as well as regulating the diameter of blood vessels and respiratory passageways.

Let us further explore these types of muscle tissue.

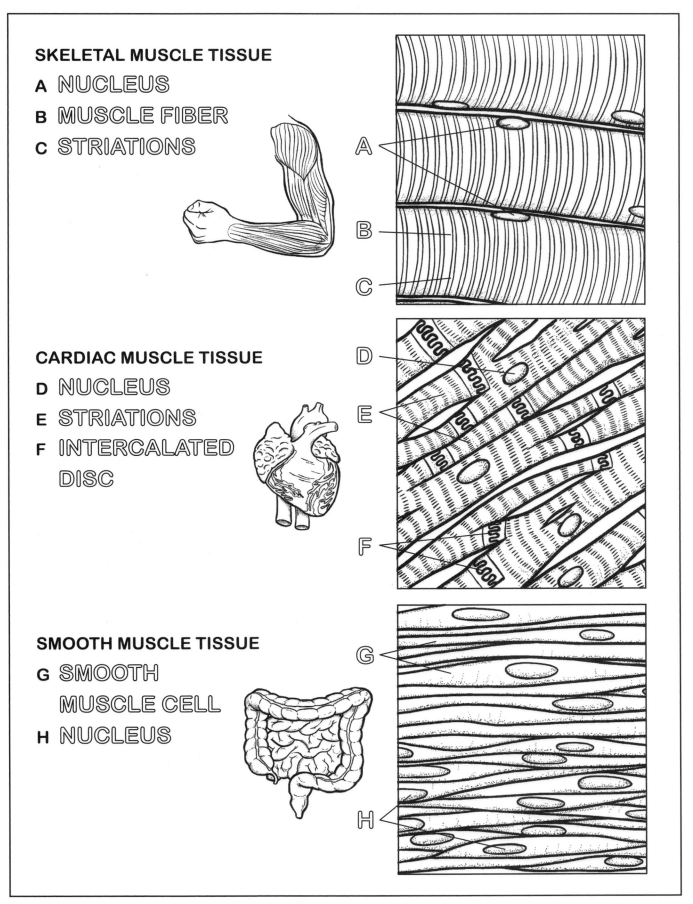

SKELETAL MUSCLE TISSUE

A NUCLEUS

B MUSCLE FIBER

C STRIATIONS

CARDIAC MUSCLE TISSUE

D NUCLEUS

E STRIATIONS

F INTERCALATED DISC

SMOOTH MUSCLE TISSUE

G SMOOTH MUSCLE CELL

H NUCLEUS

CONNECTIVE TISSUE
SKELETAL MUSCLE

Skeletal muscle fills out the body and provides it with form. You might say that this is its passive function—its active function is to move bones around joints. The unit of structure of skeletal muscle is the muscle cell or fiber.

Each muscle fiber is composed of:
• A limiting membrane: the **sarcolemma** (L., *sarco*, flesh).
• A cytoplasmic matrix: the **sarcoplasm.**
• Common cell organelles as well as the highly specialized sarcoplasmic reticulum and contractile elements, the **myofibrils.**
• Numerous **nuclei.**

Myofibrils are responsible for the longitudinal **striations** seen in skeletal muscle and are composed of myofilaments, which create the cross striations seen with the electron microscope. The functional unit of the myofilaments is a sarcomere. These myofilaments slide back and forth during muscular contraction and relaxation, and this action results in a shortening of the sarcomere. When this action is manifested throughout the whole muscle, the muscle shortens or **contracts** by about a third of its resting length.

Fibers in muscle tissue the size of the muscles of the arm, for instance, would fall apart like strings of spaghetti if they were not intimately bound together by connective tissue. By way of this connective tissue "packing," the muscle fibers receive their nerve and blood supply. Muscle tissue incorporating large amounts of this connective tissue is tough. In beef, it is this that makes certain cuts difficult to chew. Expensive meats you can cut with a fork, for example, have a reduced amount of connective tissue.

Each skeletal muscle fiber is delicately ensheathed in a thin film of reticular tissue termed **endomysium.** The endomysium blends with the **sarcolemma.**

A variable number of muscle fibers, so ensheathed, form a macroscopically visible bundle (**fasciculus**). Each fasciculus is bound by a layer of connective tissue called **perimysium**, which has continuity with the interlacing **endomysium.**

A variable number of fasciculi or skeletal muscles are bound together by a thick envelope of connective tissue, the **epimysium**, which is continuous with the finer fibers binding fasciculi and muscle fibers.

Skeletal muscles cannot contract unless they are **innervated**, or served by nerve fibers. A muscle receives a nerve from the spinal cord. The area of a skeletal muscle fiber receiving a terminal nerve fiber is called a motor end plate or myoneural junction. The nerve fiber and the muscle fibers that it innervates constitute a motor unit. We will learn more about this on page 56.

In the event of injury, muscle cells can regenerate, but slowly. Muscle cells are much more capable of enlarging (**hypertrophy**) as seen from postnatal growth or prolonged physical conditioning or exercising, or becoming smaller (**atrophying**) from disuse as seen in bedridden patients without benefit of physical therapy.

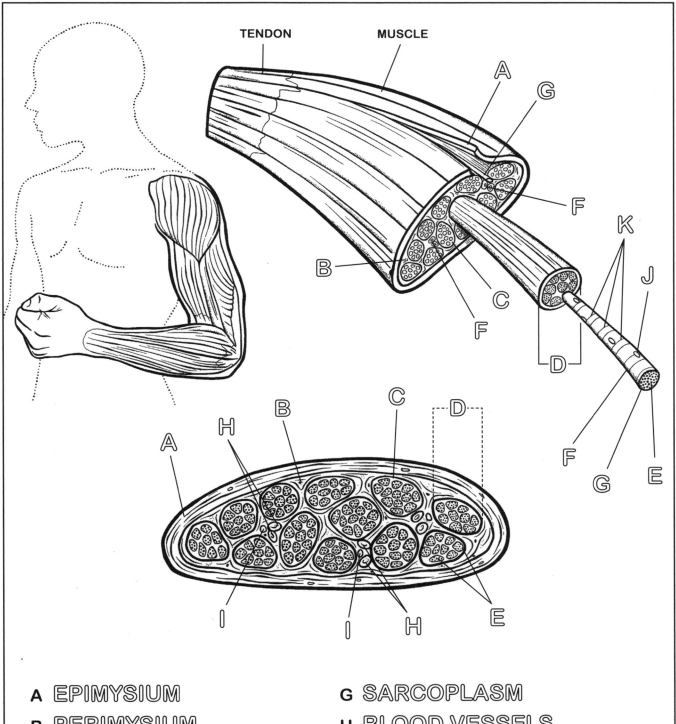

TENDON MUSCLE

A **EPIMYSIUM**

B **PERIMYSIUM**

C **ENDOMYSIUM**

D **FASCICULUS**

E **MYOFIBRIL**

F **SARCOLEMMA**

G **SARCOPLASM**

H **BLOOD VESSELS**

I **NERVE**

J **NUCLEUS**

K **STRIATIONS**

CONNECTIVE TISSUE
SMOOTH MUSCLE

Smooth muscle is visceral muscle over which you have no voluntary functional control. It is innervated by a part of the nervous system whose functioning is also independent of your will. Smooth muscle is found in the walls of the digestive tract from esophagus to anus; in the walls of the urinary tract from kidney to urethra; in the walls of the reproductive tract from uterine tube to vagina and from the sperm duct (ductus deferens) to urethra; in the walls of the respiratory tract from pharynx to respiratory tissue (alveoli); in all arteries of the body and many large veins. Smooth muscle allows you to change from far vision to near vision—it even permits you to cry! Remember the last time your hair stood on end? Smooth muscles erected those hairs.

You have no control over the movements of these organs. On the contrary, they control *you*! The **contracting** or **relaxing** of smooth muscle tells you when to eat, when to urinate, when to defecate, when to take a laxative, when to vomit, when you are going to give birth, and so on.

Smooth muscle is decidedly different from skeletal muscle on a number of counts—both histologically and functionally:

- Smooth muscle fibers are smaller than skeletal muscle fibers.
- Smooth muscle fibers do not have more than one nucleus—they are smaller than skeletal muscle fibers, and therefore do not require more than one nucleus.
- Nuclei of smooth muscle cells are centrally located in the cells; skeletal muscle nuclei are peripherally located.
- Smooth muscle fibers lack a sarcolemma and a specialized sarcoplasmic reticulum.
- Smooth muscle fibers do not each receive a nerve fiber.
- The myofibrils of smooth muscle consist of very thin myofilaments which are not well organized—hence cross striations are not apparent and thus its name, smooth muscle.
- The contractions of smooth muscle are slower but can be sustained for longer periods of time.
- Smooth muscle is more responsive to stretch.
- Smooth muscle fibers may occur singly or in small groups; skeletal muscle fibers rarely do.

Smooth muscle contraction is created by protein filaments sliding over each other. **Actin** and **myosin** are proteins found in every type of muscle tissue that work together to generate contraction and movement. In smooth muscle, these proteins are organized in a different way than other muscle cells. The actin and myosin fibers are arranged at angles to each other, connecting at **dense bodies** and at the cell membrane, stretching both across and between smooth muscle cells. Myosin acts as a tiny motor, converting molecular chemical energy into mechanical energy to slide over the actin filaments—creating shorter distances between the dense bodies and thus creating contraction and movement.

Smooth muscle may be induced into rhythmic contractions not only by nerve impulses but by hormonal stimulation—or by stretching of the muscle itself. Motor end plates are not apparent. Since each muscle fiber does not receive a nerve fiber of its own, the impulses are transmitted from one cell to another via the cell membrane.

Smooth muscle can regenerate by mitosis. Also, since smooth muscle arises from mesenchyme, in the event of injury, cells of the mesenchyme may be able to replace the injured muscle cells.

SMOOTH MUSCLE TISSUE

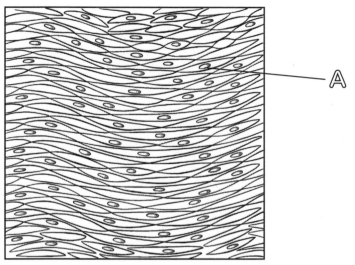

A

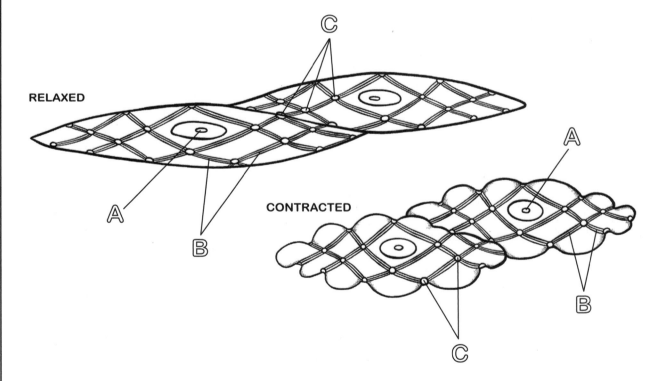

C

RELAXED

A

B

CONTRACTED

A

B

C

A NUCLEUS
B FILAMENT BUNDLES OF
ACTIN AND MYOSIN
C DENSE BODIES

CONNECTIVE TISSUE
CARDIAC MUSCLE

Cardiac muscle is the **myocardium**, the heart muscle. In several ways cardiac muscle is physiologically similar to smooth muscle and structurally similar to skeletal muscle. Let's highlight the similarities:

• Cardiac muscle, like smooth muscle, contracts rhythmically and
  the contractions are not voluntarily controlled.
• Cardiac muscle contracts spontaneously, i.e., contracts without need of innervation.
• Cardiac muscle is striated.
• Cardiac muscle has unique dark bands called intercalated discs, which
  represent the junction between adjacent cells.
• Cardiac muscle cells have single nuclei.

Cardiac muscle beats spontaneously at about forty contractions (beats) per minute, but its rate of beat (contractions) is regulated by nerves functionally independent of one's will. The myoneural junctions here are not as complex as they are in skeletal muscle. Even though cardiac muscle beats without requirement of innervation, there must be a mechanism for ensuring synchrony of contractions among all the muscle cells in the wall of each of the chambers of the heart so that the heart may pump effectively. There is, in a specialized system of modified muscle cells, the **cardiac conduction system**. It is made up of a couple of self-stimulating "electrical" impulse generators and a bundle of conducting cells (fibers) distributed through the myocardium of the heart. These fibers are rich in **glycogen**, a starch or multiple sugar, and are variably smaller and larger than cardiac muscle fibers but resemble cardiac muscle in that they have cross striations and intercalated discs.

Unlike smooth and somewhat like skeletal muscle, damaged cardiac muscle does not regenerate well at all, and following injury, the injured muscle cells, like skeletal muscle cells, will be replaced by connective tissue (scar tissue).

You might wonder how cardiac muscle can beat continuously from five weeks fetal age to seventy or even a hundred-plus years. First, the heart (cardiac muscle cells) does not beat (contract) continuously from a cellular point of view. In fact, for every second of operation of resting beating heart, each muscle fiber rests about 6/10 of that second. Second, the heart has a vast blood vessel network to provide needed oxygen. Third, the cardiac muscle fiber has a high content of glycogen from which to draw for energy.

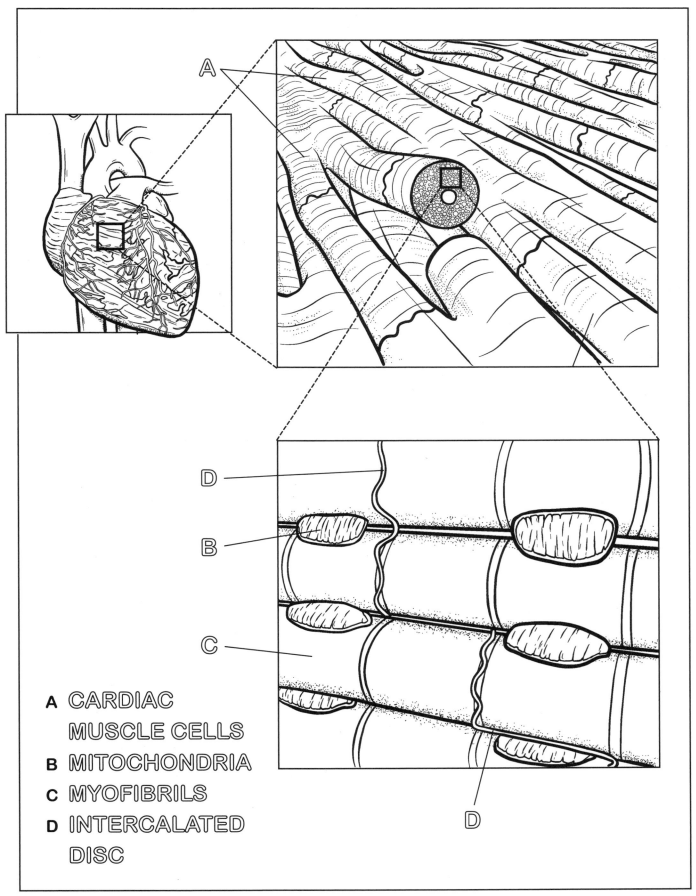

A CARDIAC
  MUSCLE CELLS
B MITOCHONDRIA
C MYOFIBRILS
D INTERCALATED
  DISC

NERVOUS TISSUE: A COMMUNICATIONS NETWORK

Nervous tissue is excitable tissue; that is, it can *generate* nerve impulses as well as *conduct* them. All of your sensations: touching, hearing, seeing, and more; all of your thoughts, emotions, memory, and reasoning; and all of your movements, from the rate of your heart's beat to the wrinkling of your nose; are based on these two phenomena.

Nervous tissue is organized into many organs whose overall function may be likened to a vast communications network. The essential function of any communication system is to receive information, and then to respond to that information in some manner. To do these things, a communication system must have:

• Receptors to sense incoming information (stimuli).
• A processing center to correlate, coordinate, integrate, and modify the information received.
• Conductors to send incoming (sensory) stimuli to the processing center, and to return evoked responses from the center to effectors which manifest the response.

Our nervous system has all these, within the following organization:

The **central nervous system (CNS)**, a processing center and central conductors, consisting of a brain and spinal cord, which process sensory information and generate appropriate responses.

The **peripheral nervous system (PNS)**, receptors and conductors outside the brain and spinal cord, consisting of spinal and cranial nerves. Some of these bring information to the CNS, and others take responses from the CNS to effector organs.

We will thoroughly explore the CNS and PNS later in this book. For now, let us learn about the cells and tissues that facilitate this communication network.

Nervous tissue is comprised of two groups of cells:

Neurons are the basic building blocks of the nervous system and receive sensory input from the external world—then transmit this information to other nerve, muscle, or gland cells. It is estimated that between 86 and 100 *billion* neurons are in the human brain!

Glial cells, also called **neuroglia**, help neurons communicate and transmit information through available communication pathways. They also work to support, nourish, and insulate neurons and neuron activity. Types of neuroglia include **oligodendrocytes**, **astrocytes**, **microglia**, and **ependymal cells**.

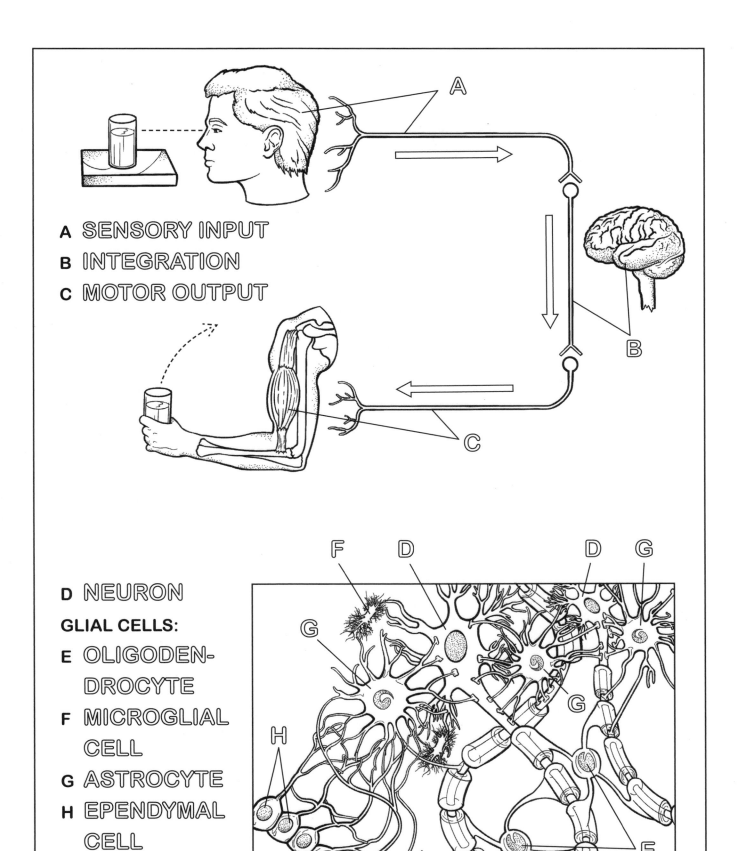

A SENSORY INPUT
B INTEGRATION
C MOTOR OUTPUT

D NEURON
GLIAL CELLS:
E OLIGODEN-
   DROCYTE
F MICROGLIAL
   CELL
G ASTROCYTE
H EPENDYMAL
   CELL

NERVOUS TISSUE
THE NEURON

The **neuron**, or nerve cell, is the fundamental unit of structure and function in the nervous system. Two components of the neuron can be identified: the **cell body**, with the nucleus and its surrounding cytoplasm; and **processes**, which are extensions of cytoplasm from the cell body.

The **cell body** (and the **nucleus**, specifically) is, of course, the administrative center for the entire neuron. Frequently, cell bodies are congregated into groups in the nervous system. In the CNS these collections of nerve cell bodies are termed **nuclei**; in the PNS these collections of nerve cell bodies are called **ganglia**. Histologically, cell bodies are characterized by:

• A large, centrally located, well-defined nucleus with prominent nucleolus.
• Blocks of the endoplasmic reticulum (ER) called Nissl substance.
• Neurofibrils and other common organelles.
• Plasma membrane.
• Inclusions (pigments, iron-containing granules, etc.), particularly in neurons of older persons.

The shape of cell bodies can vary from round to star-shaped, largely dependent upon the number of processes each has.

Neuronal processes are extensions of the plasma membrane and the cytoplasm of the cell body. They are often referred to as **nerve fibers**. These processes conduct nerve impulses to or from the cell body of origin. Neuronal processes conducting impulses toward their own cell body are called **dendrites**. These processes are short, highly branched, and receive many of the impulses coming into the cell body. They are not individually named since they do not form bundles of fibers. Their cytoplasmic constitution is like that of the cell body from which they originate.

Those processes transmitting impulses away from their own cell bodies of origin are called **axons** or axis cylinders. Whereas there may be more than one dendrite per neuron, there is only one axon per neuron. In neurons with only one process, the process is generally referred to as an axon with distal and central branches. Axons usually travel in bundles, are often long, and may give off branches. The longest axons in the human body can exceed one meter in length!

Collections of nerve cell processes of mostly axons within the CNS are called **tracts**, while collections of nerve cell processes (axons) in the PNS are called **nerves**. Most of these collections are named, e.g., optic tract, sciatic nerve.

NERVE FIBER COVERINGS
Most neuronal processes (axons) of significant size have one or more coverings, which they gather around themselves like cloaks. The sheaths are of two kinds:

Myelin is a thick, fatty sheath found in the CNS and PNS. The tiny constrictions that can be seen along a myelinated nerve fiber are places where the myelin is interrupted and the axon is laid almost bare. These are **nodes of Ranvier** and serve to facilitate the transmission of nervous impulses. Myelin is secreted by oligodendrocyte cells in the CNS and by Schwann cells in the PNS, as shown here.

The **neurilemma**, or **sheath of Schwann**, is found on most fibers of the PNS. The neurilemma consists of cells (Schwann cells) wrapped around myelinated or unmyelinated axons. The neurilemma is not found in the CNS since Schwann cells do not exist there. Neurilemma is apparently necessary for nerve regeneration, acting as a lattice for the growing nerve fibers. Hence CNS fibers cannot regenerate to any significant degree.

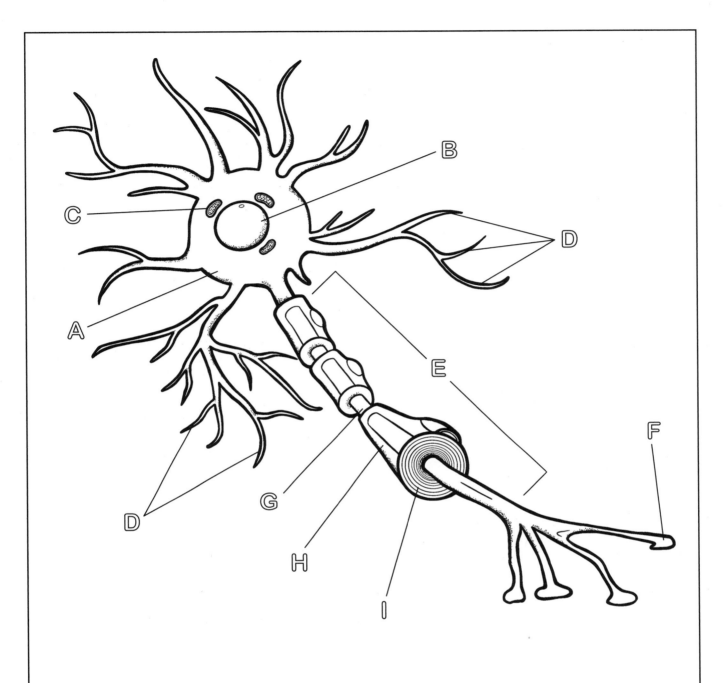

A CELL BODY
B NUCLEUS
C MITOCHONDRION
D DENDRITE
E AXON

F AXON TERMINAL
G NODE OF RANVIER
H SCHWANN CELL
I MYELIN SHEATH

NERVOUS TISSUE
NEURON CLASSIFICATIONS

Neurons may be functionally segregated into one of three classifications:

Sensory neurons, whose axons conduct information from **receptors** (distal end of the neuron near the surface of the body, in viscera, or in muscles and blood vessels), to the CNS. Their cell bodies are usually located in sensory ganglia alongside (but outside) the CNS. The processes of sensory neurons are often termed **afferent fibers**. Generally, they belong to the PNS.

Motor neurons, whose axons conduct movement-producing or gland-secreting impulses from the CNS to effector organs. Their cell bodies are located in the CNS nuclei or, in some cases, in motor ganglia outside of the CNS. Processes of motor neurons are often called **efferent fibers**. Generally, these neurons belong to the PNS.

Association neurons, or **interneurons**, which remain entirely in the CNS and form a network between the motor and sensory neurons of the PNS (and the CNS). All of the coordination, correlation, integration, and modification in the CNS occurs by a complex interrelationship of millions of association neurons.

Neurons may also be structurally classified on the basis of numbers of processes, illustrated on the adjacent page.

Unipolar neurons are sensory neurons of the PNS in which only one process (pole) extends from the cell body. Cell bodies of unipolar neurons may be seen in collections alongside the brain and spinal cord (dorsal root ganglia). The single process of this neuron splits into central and distal branches.

Bipolar neurons are sensory neurons in which two processes (one axon, one dendrite) extend from the cell body. These neurons are associated with the organs of special senses (hearing, seeing, smelling, etc.).

Multipolar neurons are those in which three or more processes (one of which is an axon) extend from the cell body. These neurons constitute a large part of the CNS and also some motor ganglia associated with the PNS. They may be motor, associative, or sensory.

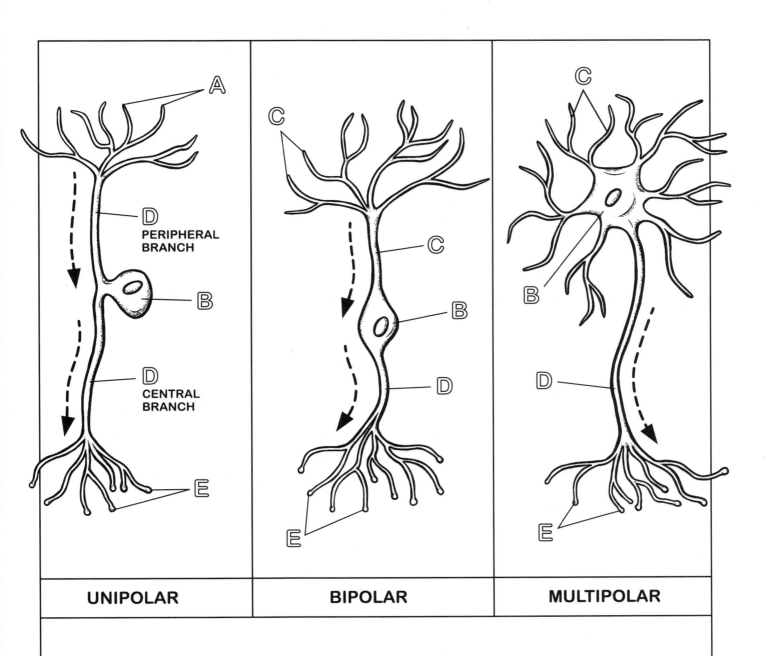

| UNIPOLAR | BIPOLAR | MULTIPOLAR |

DIRECTION OF INFORMATION

A SENSORY RECEPTOR
B CELL BODY
C DENDRITE
D AXON
E AXON TERMINAL

NERVOUS TISSUE
NERVE CONNECTIONS & ENDINGS

It has been previously stated that two of the functions of the nervous system are:

• To correlate, coordinate, integrate, and modify information received.
• To evoke appropriate responses.

To do these things, there must be a free flow of information through innumerable circuits of neurons, and this implies that neurons must interconnect. The connection between two neurons is called a **synapse**. The number of synapses is as great as the brain is complex—in fact, the complexity and efficiency of neuron communication is completely dependent upon the number of synapses. In short, the nervous system, like a politician, "has connections." The following are some characteristics of these connections:

• One neuron may receive and make more than 1,500 synapses!
• There is no structural contact between the two neurons involved in a synapse.
• The intermediary between two neurons in a synapse is chemical.
• The function of the chemical intermediary (transmitter) in a synapse is to excite the second neuron to generate an impulse or to inhibit its generation of an impulse.

NERVE CELL PROCESSES
Nerve cell processes may end in several ways:

• As a **terminal bouton**, the expanded end of an axon that synapses with the dendrite, cell body, or axon of another neuron. Such synaptic endings are found in the CNS and the ganglia of the PNS.
• As a **motor nerve ending**, the termination of a motor nerve fiber on a muscle or gland.
• As a sensory nerve ending **receptor**, the distal termination of a sensory nerve fiber; receptors are specialized for such modalities as pressure, touch, pain, heat, cold, vision, hearing, equilibrium, smell, taste, muscle sense, CO_2 concentration, and blood pressure.
• As **neuromuscular spindles**, specialized muscle fibers that act as sensory receptor complexes in skeletal muscle—the receptors fire in response to muscle stretch.

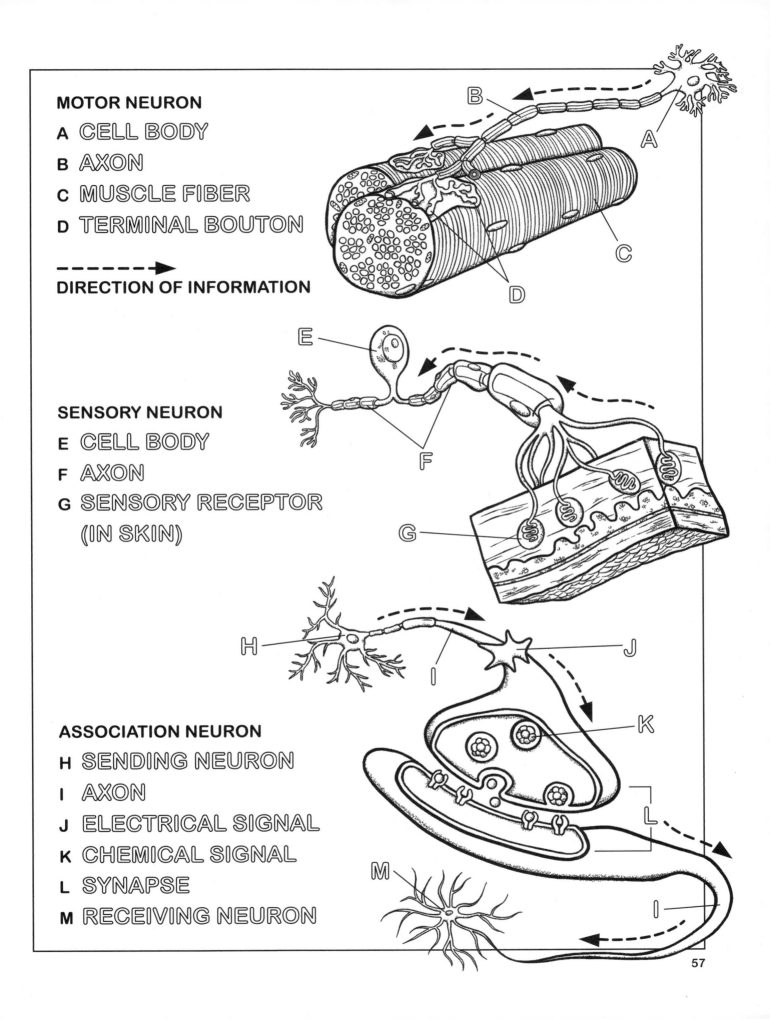

MOTOR NEURON

A CELL BODY
B AXON
C MUSCLE FIBER
D TERMINAL BOUTON

‑ ‑ ‑ ‑ ‑ ‑ ▶

DIRECTION OF INFORMATION

SENSORY NEURON

E CELL BODY
F AXON
G SENSORY RECEPTOR
(IN SKIN)

ASSOCIATION NEURON

H SENDING NEURON
I AXON
J ELECTRICAL SIGNAL
K CHEMICAL SIGNAL
L SYNAPSE
M RECEIVING NEURON

TISSUE REPAIR
& WOUND HEALING

Connective tissue plays an active part in **wound healing**. The spectrum of wounds range from sterile surgical incisions to traumatic damage to cartilage and the fracturing of bone. The discussion here will be limited to common flesh wounds (initiated by punctures and lacerations) in which the subcutaneous tissue is involved, specifically one caused by a splinter passing through the skin into the underlying connective tissue.

Wound healing involves three basic phases.

Phase one is the **hemostasis phase**: a tear through the epithelial membrane to the **subcutaneous (loose) connective tissue** by the splinter. During this phase blood and lymph (fluid and cells) leak to the outside of the wound and are visible to the naked eye. Almost immediately, a **clot** (of fibrin and red and white blood cells) is formed, preventing further fluid loss and leakage. Depending on the condition of the splinter and the flesh through which it passed, variable numbers of **bacteria** start to proliferate about the clot. Cells damaged in the initial tear release inflammatory **chemical signals**, alerting and activating pro-inflammatory cells such as **neutrophils**.

In the second **defensive/inflammatory phase**, due to the toxins released by the bacteria, acute inflammation can develop within twenty-four hours. The **capillaries** dilate, making the surface of the skin around the wound red and warm; the fluid of the blood (plasma) diffuses out into the connective tissue, causing swelling around the wound. Neutrophils enter the wound to destroy bacteria and remove debris. Pain can accompany this phase.

After around forty-eight hours, **phase three**, the **proliferative phase**, begins. Now that the clot is cleared, **fibroblasts** migrate into the vacated area and start increasing their rates of protein synthesis. Changes in the connective tissue can be noted: Cells enter into **mitosis**, a process of cell reproduction, re-covering the wound area in **epithelium**. Within one week, increased numbers of reticular fibers, and then collagenous fibers, can be seen at the periphery of the wound area. By two weeks, a thick mass of **collagenous fibers** (dense, irregular connective tissue) solidly fills the area vacated by the clot. This mass is called **scar tissue**. While the scar tissue is forming, new blood vessels are formed (by sprouting from existing capillaries) and new capillary loops are formed within a few days. Within the developing scar tissue, though, capillaries are either pushed away or do not develop at all and the scar becomes generally avascular.

Thus, a reddish wound becomes a white scar, visible to the naked eye. The tissue continues to remodel, collagen fibers reorganize, and the wound area slowly regains strength and flexibility.

As you can see, under normal conditions, the growth of connective tissue is very rapid—more rapid than any other tissue. When wounds cause large epithelial tears, the proliferating connective tissue may invade the epithelial defect and replace the epithelium.

Unlike connective tissue, nerve tissue does not regenerate well at all, except in certain circumstances. For regeneration to occur, a pathway would have to be available between the two cut ends of the nerve so they may be approximated, repaired, and reattached. In brain damage or spinal cord injuries, where the neurons are damaged in an accident, regeneration is often challenging and can often result in partial or total paralysis.

1. HOMEOSTASIS

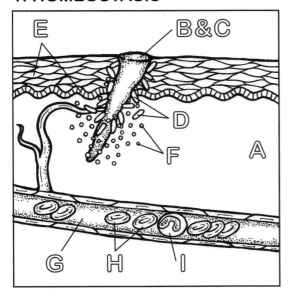

2. DEFENSIVE/INFLAMMATORY

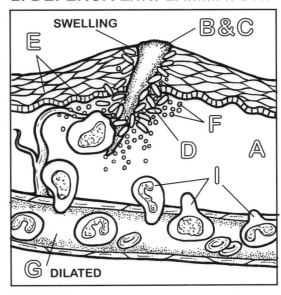

3. PROLIFERATIVE

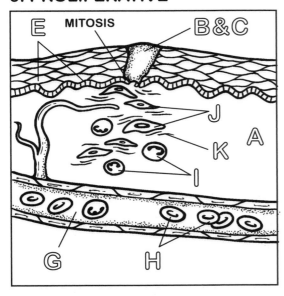

4. REMODELING

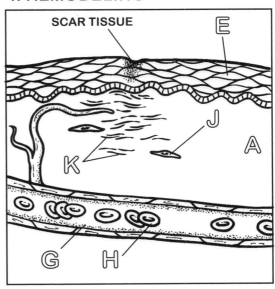

A LOOSE CONNECTIVE TISSUE

B WOUND

C CLOT

D BACTERIA

E EPITHELIUM

F CHEMICAL SIGNALS

G CAPILLARY

H RED BLOOD CELLS

I NEUTROPHILS

J FIBROBLASTS

K COLLAGENOUS FIBERS

SECTION THREE:
AN OVERVIEW OF
REGIONAL ANATOMY

The goal of this section is to prepare you for the study of the body as a functional complex of interrelated systems, organs, and tissues. Here you will get an overview of some structural and functional characteristics of bones, muscles, nerves, blood vessels, lymphatics, and the skin. This overview should make your regional study more meaningful.

Generally, the basic support of the body consists of bone, to which many, many muscles are attached, providing the body with form. Since the various bones are connected in such a way that movement between most bones is permitted, it is convenient that muscles (which contract in response to nerve stimulation) cross the connecting points (joints), enabling bone movement. We now know bone and muscle are living things, made up of collections of cells and their secretory products.

These living things need oxygen and nourishment, and receive them by way of the blood, which is conducted from place to place in living, metabolically active vessels. Lymphatic vessels assist the blood vessels in maintaining the proper balance between ions and fluids in the body tissues. The skin and underlying fascia protect the body from the elements, house the sensory receptors, and, finally, give the body a quality of smoothness.

We will study the salient features of all of these things.

INTRODUCTION TO OSTEOLOGY: BONES OF THE AXIAL SKELETON

The bones of the body make up a composite structure called the **skeleton**. The bones of the skeleton serve at least one of a number of potential functions. Such functions include:

• Support of the body, or some part of it.
• Attachment points for muscles and tendons.
• Protection for internal viscera.
• **Hemopoietic** (blood-forming) activity within the marrow cavity of long bones and others.
• A reservoir of calcium for the fluids and cells of the body (mineral content of bone is 65 percent by weight).

This internal framework of the body can be divided into the **axial skeleton** and the **appendicular skeleton**. Those bones oriented along the midline axis of the body constitute the axial skeleton, while those associated with the limbs or appendages constitute the appendicular skeleton.

Seen here, the **axial skeleton** includes:

SKULL
• Bones encasing the brain: **cranial bones** (8).
• Bones of the face: **facial bones** (14).
• Bones of the middle ear cavity: **ossicles** (6).
• Bone at the base of the tongue: **hyoid** (1).

VERTEBRAL COLUMN (BACKBONE)
• Supporting head and neck: **cervical vertebrae** (7).
• Supporting chest, neck, and head; articulates with ribs: **thoracic vertebrae** (12).
• Supporting abdomen, ribs, and chest thorax; neck and head: **lumbar vertebrae** (5).
• Supporting the weight of the body less the lower limbs; articulates with and supports the hip bone: **sacrum** (five fused bones).
• The vestigial tail: **coccyx** (four fused bones).

THORAX
• The breastplate: **sternum** (1).
• Bones that surround the thoracic viscera and some of the abdominal viscera as well: **ribs** (24).

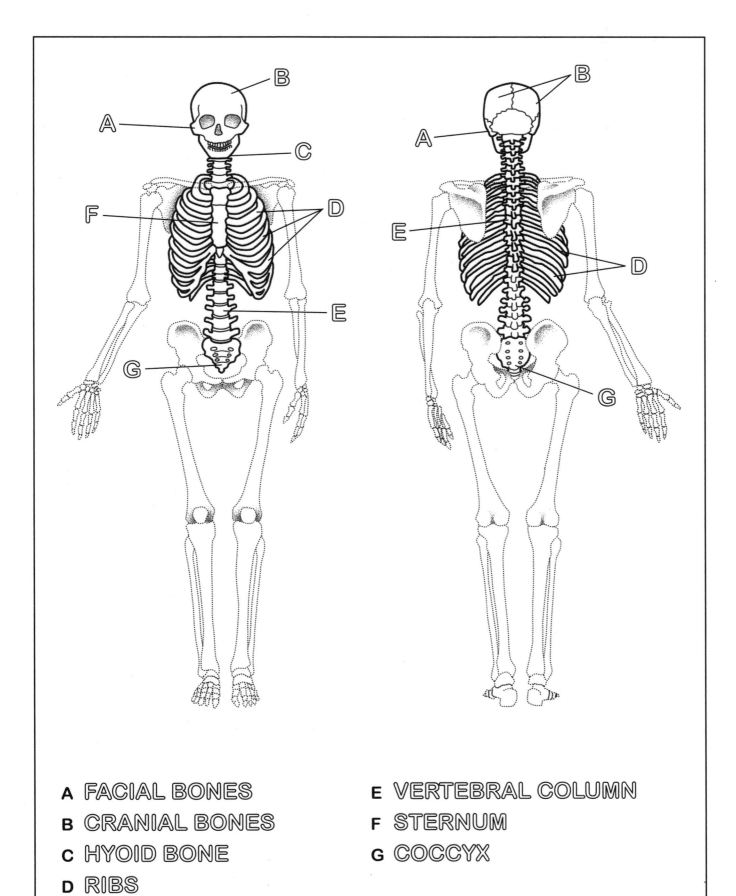

A FACIAL BONES

B CRANIAL BONES

C HYOID BONE

D RIBS

E VERTEBRAL COLUMN

F STERNUM

G COCCYX

BONES OF THE APPENDICULAR SKELETON

The **appendicular skeleton** consists of a pair of bony girdle and the limb bones to which they connect. In the case of the upper limb, the girdle surrounds the upper chest (pectoris) and is conveniently termed the **pectoral girdle**. The members of the girdle include:

• The anterior components: **clavicle**(s).
• The posterior components: **scapula**(e).

From this scapulae, the limb bones are fastened in the following sequence:

• Bones of the arms: **humerus**.
• Bones of the forearms: **ulna** and **radius**.
• Bones of the wrists: **carpals**.
• Bones of the hands and fingers: **metacarpals** and **phalanges**.

The bones of the lower limb include the limb bones and the girdle of bone (the hip bones) encircling and thereby forming the pelvis—conveniently termed the **pelvic girdle**. From the hip bones, articulated in sequence, are:

• Bones of the thighs: **femur**.
• Bones of the legs: **tibia** and **fibula**.
• Bones of the feet: **tarsals**, **metatarsals**, and **phalanges**.

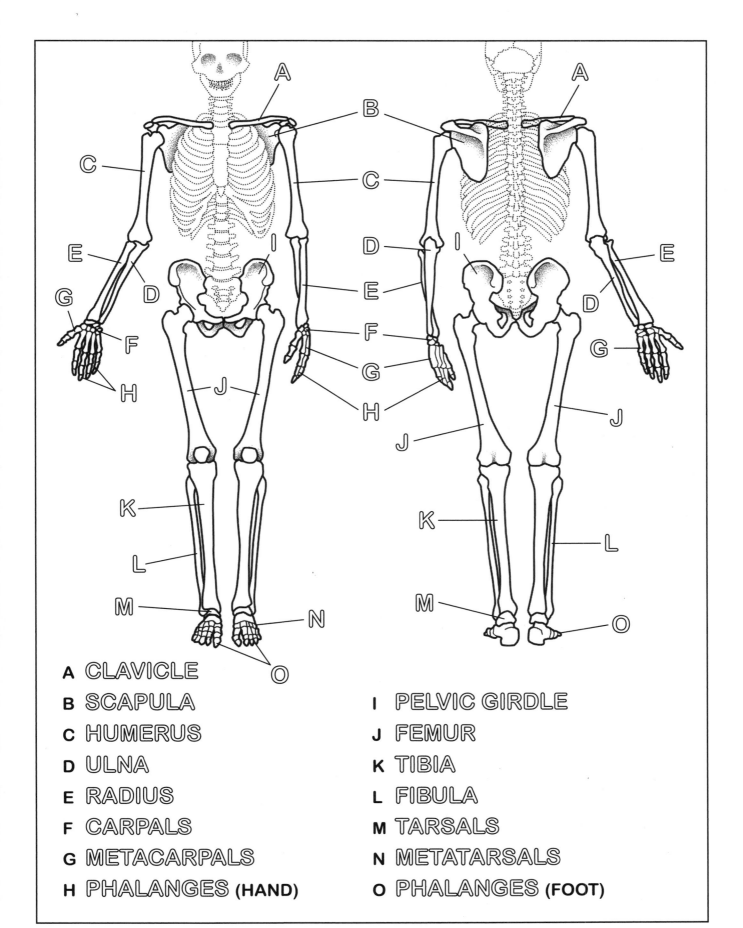

A CLAVICLE
B SCAPULA
C HUMERUS
D ULNA
E RADIUS
F CARPALS
G METACARPALS
H PHALANGES (HAND)

I PELVIC GIRDLE
J FEMUR
K TIBIA
L FIBULA
M TARSALS
N METATARSALS
O PHALANGES (FOOT)

CLASSIFICATION OF BONES ACCORDING TO SHAPE

All the bones of the skeleton, once sorted out, generally fall into four classes according to shape.

Long bones: Those with a long axis, or more length than width.

Short bones: Those that are stubby, more or less cube-shaped.

Flat bones: Plate-like bones.

Irregular bones: All other bones.

Among irregular bones is also a group of small, round **sesamoid bones** which develop in the tendons of certain muscles. Their name derives from the Latin word *sesamum* ("sesame seed"), indicating the size of most sesamoids. Found in the foot, hand, and the knee, these bones apparently formed, in evolutionary terms, in response to friction generated by the tendons passing over bone. The largest and most significant sesamoid bone in our body is the **patella**—the kneecap.

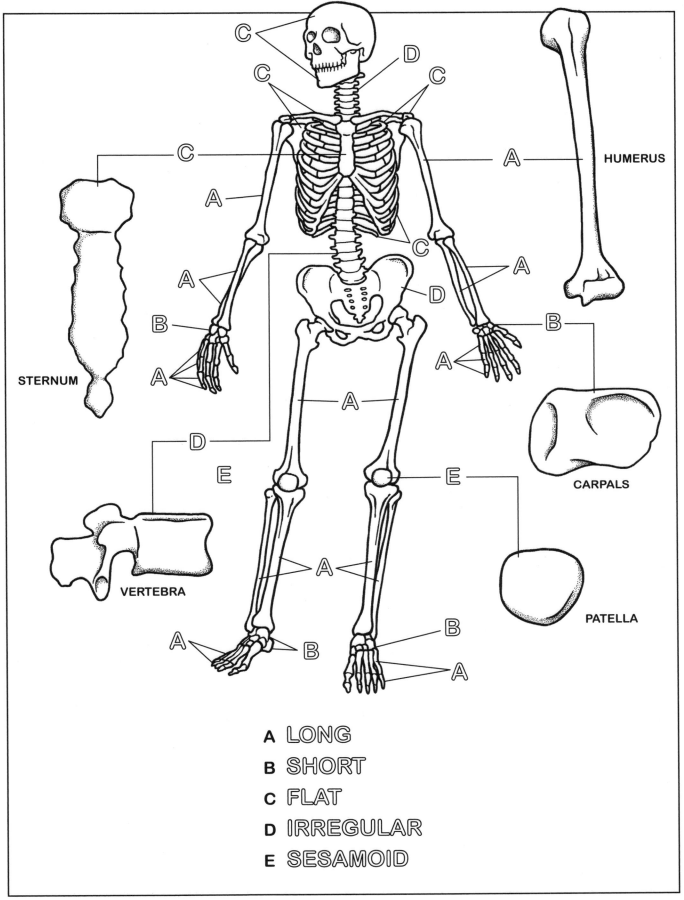

STERNUM

HUMERUS

CARPALS

VERTEBRA

PATELLA

A LONG
B SHORT
C FLAT
D IRREGULAR
E SESAMOID

CHARACTERISTICS OF BONE SHAPE

During formation and growth, bones develop holes, projections, and depressions. The holes, or **foramina** (sing., **foramen**) are created by the passage of blood vessels or nerves into or out of the bone. Ridges and grooves of bone are often created by nerves and/or vessels pressing on the soft developing bone (much as a paw print is formed in newly laid cement). Many of the projections and depressions of bone are created by the pull of muscles and tendons (so-called **traction epiphyses**).

The terminology for the various foramina, ridges, grooves, projections, and depressions involves the frequent use of synonyms. For example:

A hole in a bone:
- If round, is a **foramen**,
- If slit-like, is a **fissure**,
- If tube-like, is a **canal** or **meatus**,
- If hollow space, is a **cavity**.

A ridge on a bone:
- If extensive, may be called a **crest**,
- If thin, may be called a **line**.

A groove on a bone (a depression with a long axis):
- May be termed a **sulcus** or **fissure**.

A projection on a bone:
- Of large proportions, may be a **trochanter**, a **tuberosity**, or a **malleolus**,
- If small, may be a **tubercle** or **process**,
- If pointed or spear-like, may be a **spine** or a **styloid process**.

A depression in a bone:
- If small, is a **fovea**,
- If larger and deeper, is a **fossa**,
- If moon-shaped, may be a **notch**.

Whenever a bone articulates with another, the articular surface, capped with hyaline cartilage:
- If rounded, a **condyle** or **head**,
- If flat, a **facet**.

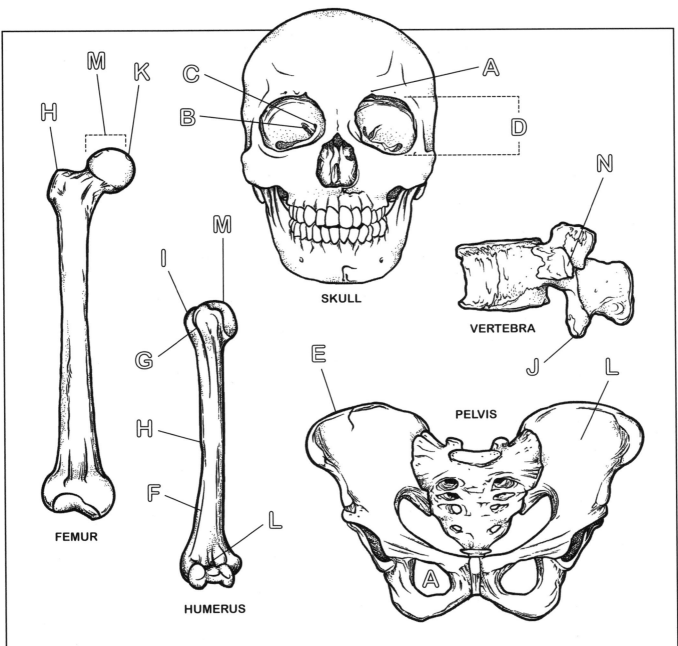

SKULL

VERTEBRA

FEMUR

HUMERUS

PELVIS

A FORAMEN

B FISSURE

C CANAL/MEATUS

D CAVITY

E CREST

F LINE

G SULCUS/FISSURE

H TROCHANTER/
TUBEROSITY

I TUBURCLE/PROCESS

J SPINE

K FOVEA

L FOSSA

M CONDYLE/HEAD

N FACET

INTRODUCTION TO ARTHROLOGY: FIBROUS & CARTILAGINOUS JOINTS

An **articulation** describes the joining of two or more bones. There are various kinds of joints, classified according to type of movement permitted or what the joint is made of.

• Functionally, joints may be **immovable**, **partly movable**, or **freely movable**.

• Structurally, joints are **fibrous**, **cartilaginous**, or **synovial**.

Fibrous and cartilaginous joints may be immovable or partly movable, but they are never freely movable.

Fibrous joints are those in which the intervening substance between adjoining bones is fibrous (dense regular or irregular) connective tissue. Such joints may be seen between bones of the skull; these joints are often called **sutures**. With aging, the fibrous tissue is replaced by bone and the suture becomes **synostosis**, an immovable joint. Fibrous joints are also seen between two bones that are some distance apart. The intervening tissue is regularly arranged dense connective tissue (ligaments). Seen between the bones of the forearm and the bones of the leg, a joint of this type is often termed *syndesmosis* and described as an **interosseous membrane**. Syndesmoses are slightly movable.

Cartilaginous joints are those in which the intervening substance between adjoining bones is cartilage. Such joints may be seen between the epiphysis and the diaphysis of developing bone. You will remember seeing these before. They were called epiphyseal plates. Recalling the descriptive terms synostosis and syndesmosis, how would you describe the above joint in one word? *Synchondrosis*. Such a joint is immovable and in young adulthood becomes a synostosis. Cartilaginous joints are also found between bones where the intervening substance is a fibrocartilage (in addition to the hyaline cartilage over the articular surfaces). This joint is slightly movable and may be seen at the interpubic joint, intervertebral joint, and between upper segments of the sternum. Such joints are called *symphyses*.

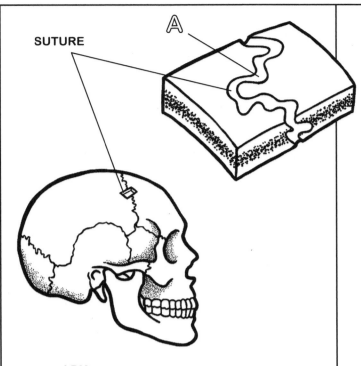

SUTURE

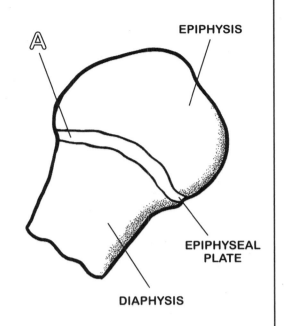

EPIPHYSIS

A

EPIPHYSEAL
PLATE

DIAPHYSIS

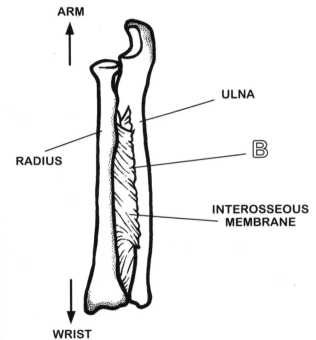

ARM

ULNA

B

RADIUS

INTEROSSEOUS
MEMBRANE

WRIST

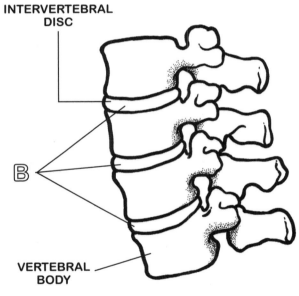

INTERVERTEBRAL
DISC

B

VERTEBRAL
BODY

FIBROUS JOINT

A IMMOVABLE

B PARTLY MOVABLE

CARTILAGINOUS JOINT

A IMMOVABLE

B PARTLY MOVABLE

SYNOVIAL JOINTS

Synovial joints are freely movable. They are also characterized by:

• A fibrous articular capsule enclosing the two articular surfaces, thus forming a **joint cavity**. These capsules, often reinforced by ligaments, arise from and merge with the periosteum on the sides of the bone. The capsules may be strong (as in the hip joint) or weak (as in the shoulder joint).

• A **synovial membrane** lining the interior of the fibrous capsule. This loose connective tissue membrane (epithelial) forms folds within the joint cavity but does not cover the actual articulating surfaces.

• A **synovial cavity** with fluid secreted by the synovial membrane. It is a watery fluid, much like (and may be) plasma and acts to lubricate the joints and prevent overheating.

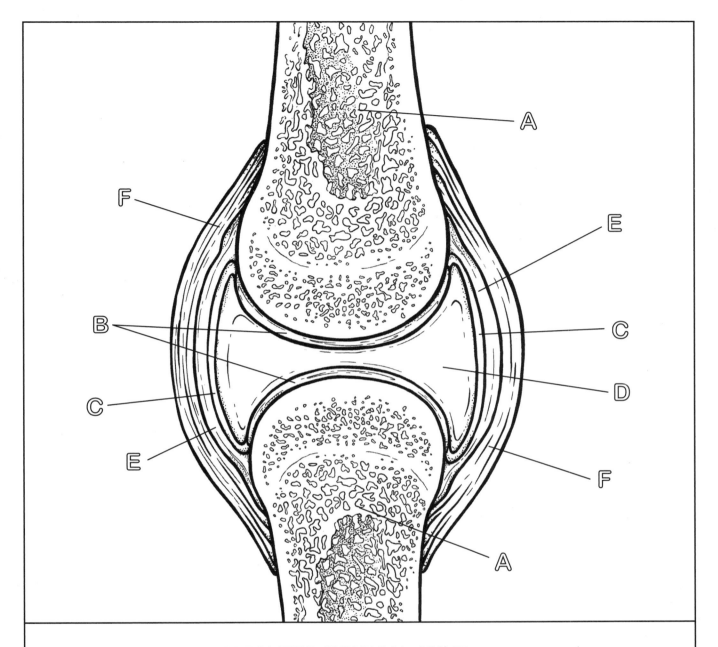

IDEALIZED SYNOVIAL JOINT

A ARTICULATING BONES
B ARTICULAR CARTILAGE
C SYNOVIAL MEMBRANE
D SYNOVIAL CAVITY (FLUID)
E JOINT CAVITY
F COLLATERAL LIGAMENT

TYPES OF
SYNOVIAL JOINTS

Synovial joints may also incorporate other structures that are necessary to joint stability. Such structures include articular discs or menisci, pads of fibrocartilage, rings of fibrocartilage around sockets (thus deepening the socket), or extra ligaments reinforcing the capsule.

Synovial joints may be one of several varieties. Specific joints will be considered with the appropriate region of the body. Some examples of synovial joints are:

• **Ball-and-socket joints** (multiaxial): Characterized by articular surfaces shaped like a ball and a socket. Rotation and circumduction are allowed.

• **Condylar joints**: Where one bone is a socket and the other fits into the socket but the movements are limited to **circumduction** (no rotation). This is really a lesser degree of a ball-and-socket joint.

• **Pivot joints** (also uniaxial): Where one bone rotates on or about another, much as a wheel rotates around its axle.

• **Gliding joints** (or plane joint): Where the two articular surfaces are flat and slide across one another.

• **Hinge joints**: Where one articular surface is concave and the other is convex. Movement is limited to one axis. In terms of movement, this is a uniaxial joint.

• **Saddle joints** (biaxial): Where the concavity of one surface moves in the concavity of the other.

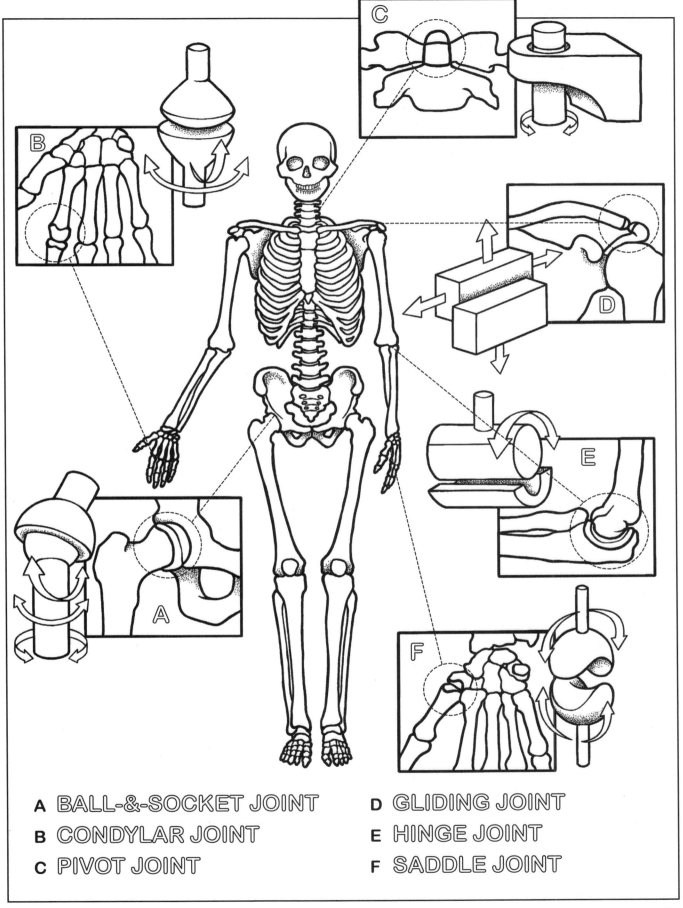

A BALL-&-SOCKET JOINT
B CONDYLAR JOINT
C PIVOT JOINT
D GLIDING JOINT
E HINGE JOINT
F SADDLE JOINT

INTRODUCTION TO SKELETAL MUSCLE: MYOLOGY

A skeletal muscle generally has two parts: the contractile unit—the **belly**—and the ends that are attached to bones—**tendons**. Muscle fibers themselves generally do not attach directly to bones. A tendon attaches to bone by merging with its **periosteum** or by inserting directly into the bony tissue. In some cases, such as in the muscles of the face, tendons insert into and merge with the superficial fascia underlying the skin. If the tendon is flat and sheet-like in appearance, it is an **aponeurosis**.

The individual skeletal muscle fibers do not generally extend throughout the whole length of a muscle. The muscle fibers are connected to fellow parallel fibers by connective tissue (**endomysium**).

A muscle usually is attached (by tendons) to two bones that abut at a joint; one of those bones will move away from or toward the other end when the muscle contracts, depending on the joint involved and the muscle's relation to that joint. It is important to differentiate the two attachments in order to appreciate the muscle's function:

• The attachment to the more fixed or nonmoving bone is the **origin**.

• The attachment to the moving bone is the **insertion**.

The arrangement of fibers in skeletal muscle is variable but is frequently related to the muscle's functional capacity. The variation of fiber architecture is dependent on orientation of the tendons, presence of connective tissue septa, length of muscle, and so on.

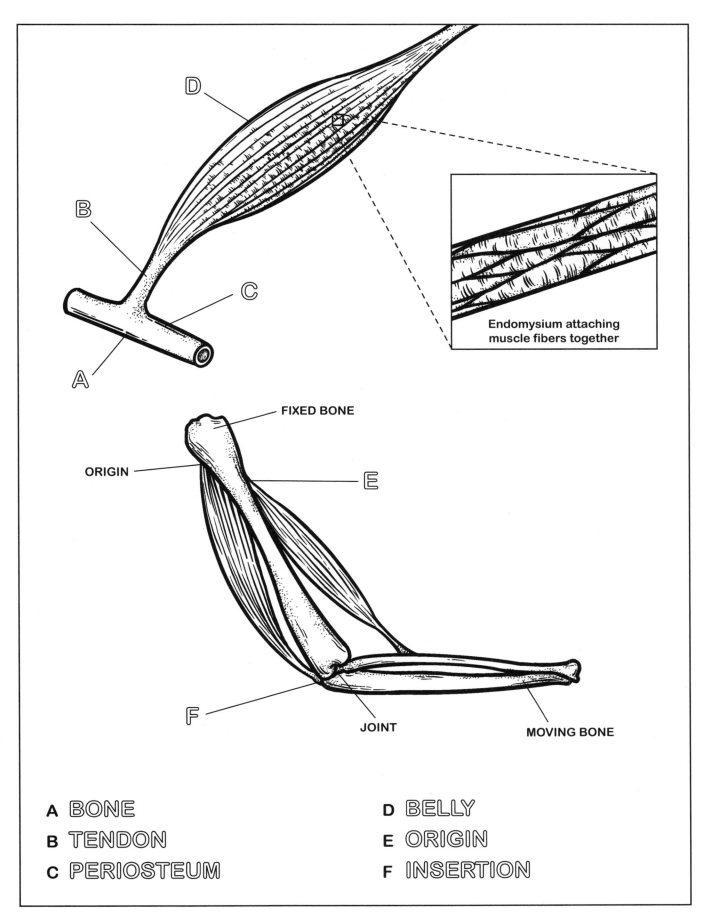

D

B

C

A

Endomysium attaching
muscle fibers together

FIXED BONE

ORIGIN

E

F

JOINT

MOVING BONE

A BONE

B TENDON

C PERIOSTEUM

D BELLY

E ORIGIN

F INSERTION

HOW MUSCLES WORK: SIMPLE MACHINES OF THE MUSCULOSKELETAL SYSTEM

To appreciate the mechanics of skeletal muscle action, it is necessary to consider some basic mechanical principles. In order to act effectively, muscles employ the concept of mechanical advantage. They do this through the utilization of the principles of simple machines. A machine is merely a contrivance that modifies forces and movements so that work can be done more efficiently. The musculoskeletal system employs three of the six universally accepted simple machines: (1) the lever, (2) the wheel and axle, and (3) the pulley.

LEVERS
Levers require three basic components:

• **Fulcrum** (F): The axis or point in space about which a structure, e.g., bone, moves or turns. In the musculoskeletal system, such a point would be a joint.

• **Load** (L): The structure, e.g., bone, that will be moved or turned about a fulcrum and that offers resistance to the force attempting to move it.

• **Power** (P): The force applied to the load so that the movement may occur. In the body, such a force would be derived from muscular work.

WHEEL AND AXLE
An example of the wheel and axle machine may be seen in the head of the atlas (first cervical vertebra) pivoting about the axis (the second cervical vertebra), much as a wheel rotates about an axle.

PULLEY
Pulleys change the direction of an applied force. In many cases, two or more muscles may act as parallel forces in carrying out a particular task. If these forces are directed oppositely, the effect is one of rotation and the forces are collectively termed a **force couple**.

The mechanics of physical activity, just superficially introduced here, can be analyzed in more detail in **kinesiology**, the scientific study of human body movement.

INTEGRATION OF MUSCLE ACTION
In general, a number of muscles or muscle groups are active in any given movement; rarely does one muscle act alone to provide movement. In fact, a high degree of integration and coordination of muscular activity is often required.

WHEEL AND AXLE

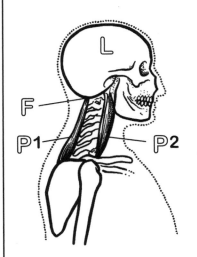
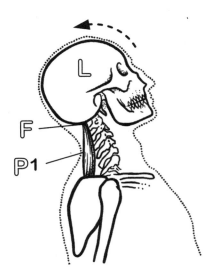
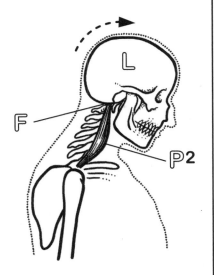

AT REST   BACKWARD MOVEMENT   FORWARD MOVEMENT

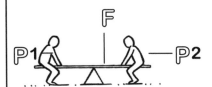

PULLEY

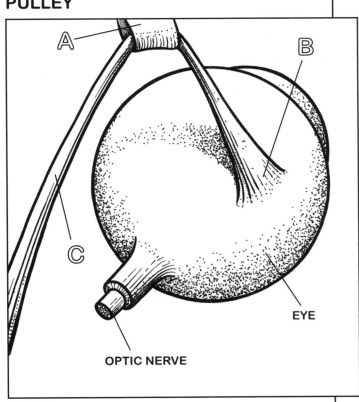

F FULCRUM
L LOAD
P₁ POWER
P₂ POWER

A PULLEY
B TENDON
C MUSCLE

EYE

OPTIC NERVE

FASCIAE & BURSAE

Fascia (pl., fasciae) is a layer of loose or dense irregular connective tissue throughout the body. In certain places it may be known by special terms (e.g., *submucosa* and *lamina propria* in viscera) but generally fascia is simply referred to as being either **superficial** or **deep**.

The fascia just under the skin, variable in thickness and density, is **superficial fascia**, and is generally described as having an outer fatty layer (**subcutaneous**) and then a **fibrous** layer. Either of these layers may be reduced or absent. The subcutaneous layers add a quality of smoothness to the form of the body. The female sex hormone estrogen influences the distribution of fat at puberty and the result is that generally a female body form appears smoother than the more angular form of the male. The fibrous layer of superficial fascia has special significance for the surgeon, for its tough fibrous quality lends itself to stitching (suturing) wounds in surgical operations.

In general, superficial fascia:

• Insulates the body against cold.
• Pads the space between skin and muscles.
• Protects transiting nerves and vessels.
• Acts as a reservoir of energy—fat from fascia is quickly mobilized when the body requires it, and is used as a source of energy for metabolic activity; thus the body form can change if these energy reservoirs are used.

Deep fascia, in the conventional sense, constitutes the connective tissue "envelope" bounding skeletal muscle (external to but contiguous with perimysium), and forming walls (**septa**) between muscles and muscle groups. It also contributes significantly to the construction of joints. Deep fascia fills out regions of deeper structures just as superficial fascia fills out the external body form. Deep vessels and nerves pass through and are protected by deep fascia. Deep fascia is often continuous with periosteum, ligaments, and superficial fascia—providing a firm network of fibrous support for the muscles, viscera, vessels, and nerves within.

Bursae (sing., bursa), like the collection shown opposite around the shoulder joint, are partially collapsed connective tissue sacs—imagine partially filled water balloons as slippery padding! Bursae are found in the superficial fascia and between tendons and bones, muscles and bones, muscles and muscles, or wherever adjacent musculoskeletal structures rub together and create the heat of friction. These bursae prevent the moving parts from inflaming one another due to contact irritation. The inner surface of the bursae is like a synovial membrane and secretes a fluid (synovia) which moistens the interior, sufficient to allow one surface to rub on another frictionlessly. Not infrequently, bursae may communicate with joint cavities following chronic inflammation of adjacent tendons (tendonitis) or the bursa itself (bursitis).

In places where tendons travel back and forth over bones, a tubular bursa, secured in place by a fibrous sheath, surrounds the tendon. Such a bursa is called a **tendon sheath**; it allows frictionless movement of the tendons through the tunnels. These sheaths are particularly significant in the hand, as untreated infections of them can result in permanent functional impairment.

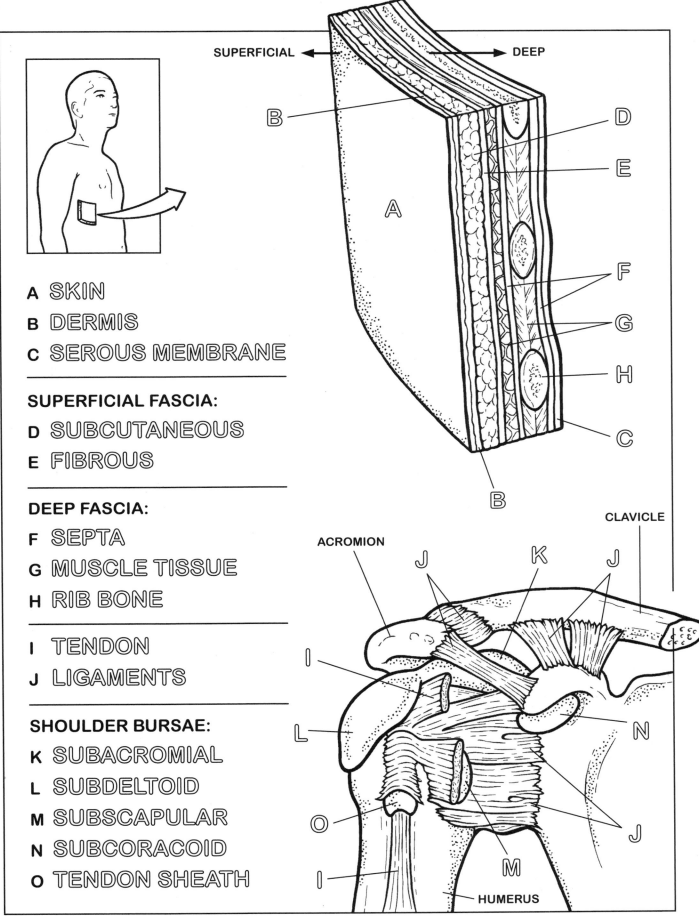

SUPERFICIAL ← → DEEP

A SKIN
B DERMIS
C SEROUS MEMBRANE

SUPERFICIAL FASCIA:

D SUBCUTANEOUS
E FIBROUS

DEEP FASCIA:

F SEPTA
G MUSCLE TISSUE
H RIB BONE

I TENDON
J LIGAMENTS

SHOULDER BURSAE:

K SUBACROMIAL
L SUBDELTOID
M SUBSCAPULAR
N SUBCORACOID
O TENDON SHEATH

CLAVICLE

ACROMION

HUMERUS

81

INTRODUCTION TO THE CARDIOVASCULAR SYSTEM

The **cardiovascular system** is composed of a pump (heart) and system of tubes (vessels) that:

• Carry blood pumped away from the heart via **arteries**;
• Distribute and receive certain molecular material, both nutritive and waste, to and from local tissues via **capillaries**;
• Carry blood to the heart from the tissues via **veins**.

The heart is explored in more detail on page 290, along with viscera of the thoracic cavity.

Within this circulatory system, one set of vessels is responsible for carrying oxygenated blood and nutritive material to the tissues and cells of the body and returning deoxygenated blood to the heart. This is the **systemic circuit**. Another set of vessels carries the deoxygenated blood from the heart to the lungs for oxygen replenishment and returns the oxygenated blood to the heart. This is the **pulmonary circuit**. Oxygenated blood is reddish in color (due to the nature of oxygen-binding hemoglobin molecules in red blood corpuscles) while deoxygenated blood is bluish (because of the reduced concentration of oxygen).

It should be appreciated that blood vessels are not simply a system of inert tubes, but are metabolically active organs composed of epithelial, connective, muscle, and nervous tissues. The structure of the types of vessels described above varies significantly and is related to the task performed. The functions of a particular vessel is related to the latter's proximity to the heart as well as to the amount of each of the tissues present in the walls. For instance, a vessel receiving blood from the heart (artery) at high pressure will certainly be structured differently from a vessel carrying blood at relatively low pressure toward the heart (vein); and a vessel composed primarily of elastic tissue certainly functions differently from a small epithelial capillary.

All these points will be addressed in the following pages.

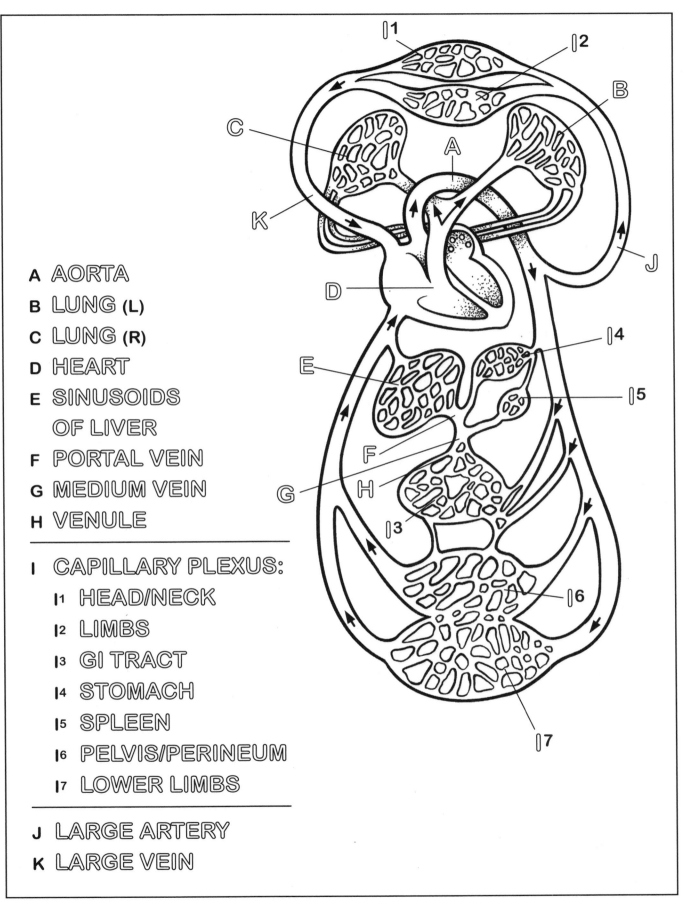

A AORTA
B LUNG (L)
C LUNG (R)
D HEART
E SINUSOIDS
  OF LIVER
F PORTAL VEIN
G MEDIUM VEIN
H VENULE

I CAPILLARY PLEXUS:
  I1 HEAD/NECK
  I2 LIMBS
  I3 GI TRACT
  I4 STOMACH
  I5 SPLEEN
  I6 PELVIS/PERINEUM
  I7 LOWER LIMBS

J LARGE ARTERY
K LARGE VEIN

THE CARDIOVASCULAR SYSTEM
ARTERIES

Those vessels carrying blood from the heart to the tissues are called **arteries**. The arteries arising directly or almost directly from the heart are referred to as **large arteries**. As they spring out of the thoracic cavity and branch into smaller regional arteries they are called **medium arteries**. The smallest branches of these, communicating with the capillary networks (through which oxygen, nutritive elements, and wastes pass into and out of the neighboring tissues) are called **arterioles** or **small arteries**.

Each of these classes of arteries differs functionally and structurally. The large arteries are responsible for conducting blood from the heart to the regional vessels without sacrificing any significant pressure loss. They do this by stretching their walls during the blood ejection phase of the heart (**systole**) and recoiling during the relaxation phase (**diastole**). What kind of tissue would you think predominates here? Elastic connective tissue! Hence, large arteries are often termed **elastic** or **conducting arteries**. Furthermore, large arteries all have small vessels perforating their walls to supply them with blood. Such vessels constitute **vasa vasorum**.

The medium-sized arteries supply blood to the various regions of the body. Thus, these arteries are often termed **distributing arteries**. Blood flow to one or more regions can be altered by varying the diameter of the appropriate distributing arteries. What kind of tissue, found in the walls of these vessels, is responsible for this action? Smooth muscle. For this reason, these arteries are also called **muscular arteries**.

The small arteries (including the **arterioles**) generally cannot be seen by the naked eye. Like muscular arteries, they contain one or more layers of smooth muscle and are responsible for regulating blood flow into the capillary networks. The greatest resistance to blood flow is produced here, and this is reflected in the blood pressure measurements of an individual, i.e., the greater the resistance, the higher the pressure. Thus arterioles, as well as muscular arteries, are referred to as **resistance vessels** by the physiologist.

These different types of blood vessels vary slightly in their structures, but they share the same general features. The walls of most blood vessels have three distinct layers called **tunics** (from the Latin term *tunica,* after the garments worn by ancient Romans): the **tunica externa**, the **tunica media**, and the **tunica intima**. These surround the hollow interior through which blood flows, known as the **lumen**. In larger arteries we have learned there is also a layer of elastic tissue known as the **internal elastic membrane** (also called the **internal elastic lamina**), poviding structure while allowing the vessel to stretch.

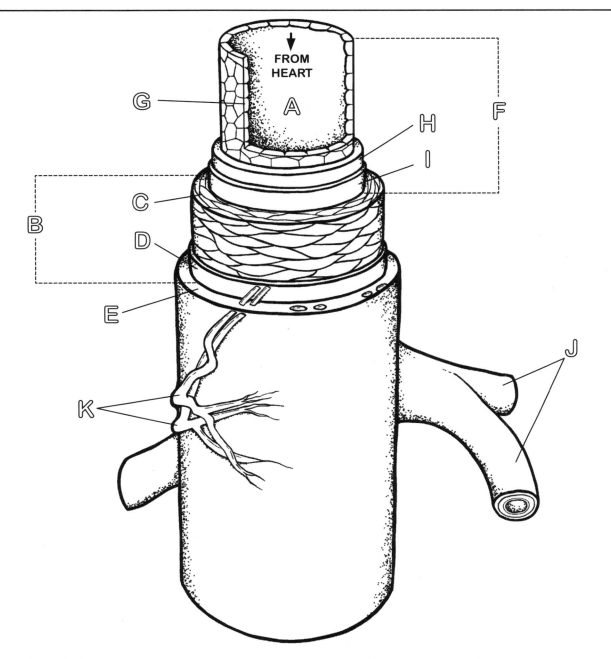

A LUMEN

B TUNICA MEDIA

C SMOOTH MUSCLE

D EXTERNAL ELASTIC
   LAMINA

E TUNICA EXTERNA

F TUNICA INTIMA

G ENDOTHELIUM

H SUBENDOTHELIUM

I INTERNAL ELASTIC
   LAMINA

J ARTERIOLES

K VASO VASORUM

THE CARDIOVASCULAR SYSTEM
VEINS

Arteries conduct blood from the heart to capillaries, from which cells of the body receive their nutrition. Waste products are discharged by the cells and taken up by the capillaries, which merge into vessels returning blood to the heart. Such vessels are **veins**. As veins get closer to the heart, they become larger, just as rivers become larger as they approach the sea. This is so because veins, like rivers, receive more and more tributaries as they near the heart, or sea. The smallest veins are called **venules**. Arising from capillaries, venules are the tributaries of medium veins. These, in turn, are tributaries of the larger veins further downstream.

It should be noted that while veins have tributaries, arteries do not. Because the direction of blood flow in arteries is away from the heart, arteries are said to have branches.

Venules are about the same size as arterioles, but the pressure of the blood within venules is about half the pressure in arterioles. This fact is reflected in their thinner walls and larger **lumen**.

Veins have the capacity to distend and thus function as a reservoir of blood and are often called **capacitance vessels** (in contrast to small and medium arteries which are called **resistance vessels**). Larger veins, like arteries, are supplied by vasa vasorum.

In medium-sized veins of the extremities and other areas, **valves** may be found. These structures function to prevent retrograde (back) flow in those veins where gravity has a significant influence. The valves, usually arranged in pairs, are cup-shaped endothelial pockets. The opening of the pockets always face the heart. Such venous valves can often be seen in the neck of people yelling or straining!

Medium-sized arteries and veins usually travel together as they pass through a body region. This is not a chance meeting, as veins and arteries develop together from the same embryonic tissue. Particularly in the extremities, medium-sized veins traveling with a companion artery (and nerve) will often travel in pairs, with one on each side of the artery—referred to as **venae comitantes** (sing., *vena comitans,* is Latin for "accompanying vein").

Certain organs of the body include enlarged capillary-like vessels termed **sinusoids** in their structure, particularly where a great deal of molecular exchange takes place. Conducting blood toward the heart, such venous sinusoids are found in the liver, bone marrow, and other "bloody" organs, and will be considered regionally. Sinusoids are often lined by phagocytic cells. Greatly enlarged veins within the skull and placenta are called **venous sinuses**.

THE CONCEPT OF COLLATERAL CIRCULATION
It should be appreciated that were it not for some sort of alternate routes of vascular circulation about the regions of the body, significant cuts and wounds involving **ligation** (tying) of distributing arteries would mean cutting off the blood supply to the part, with gangrene following. Cells must have a constant supply of oxygen (brought by blood). Interruption of that blood supply for any more than a few seconds would result in the death of many cells. Fortunately, there are usually a variety of vascular routes to any one region of the body and within that region there are interconnecting vessels. A connection between vessels is an **anastomosis**. Anastomoses provide alternate routes of blood circulation; such routes are called **collateral circulation**. Some organs are notably devoid of any significant collateral circulation, such as the liver, kidney, and heart.

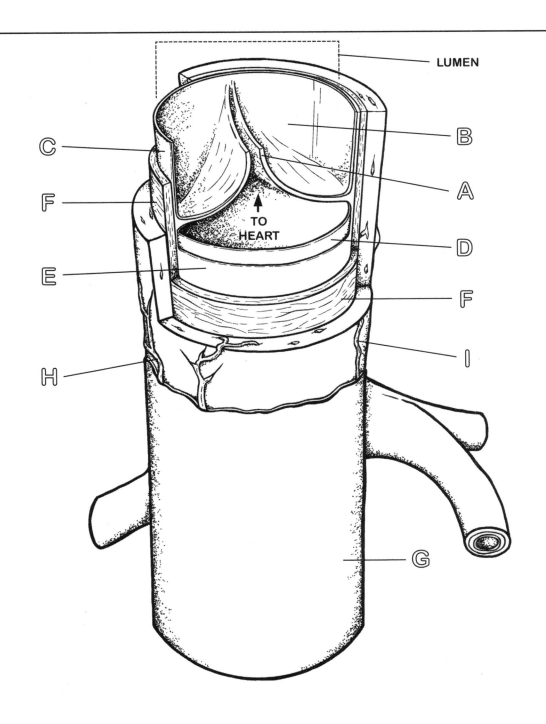

LUMEN

TO
HEART

A VALVE

B ENDOTHELIUM

C TUNICA INTIMA

D BASEMENT MEMBRANE

E SUBENDOTHELIUM

F TUNICA MEDIA

G TUNICA EXTERNA

H VASA VASORUM

I NERVI VASORUM

THE CARDIOVASCULAR SYSTEM
CAPILLARIES

Capillaries are the smallest and thinnest of all the blood vessels—and also the most common. It's been estimated that there are 40 billion capillaries in the average human body! Capillary walls are made of a thin cell layer called **endothelium** that's surrounded by another thin layer called the **basement membrane**. They are so small that red blood cells need to flow through them single file. They usually cannot be directly seen by the naked eye, but their presence is reflected in the color of the nail, the eyes, etc. When one becomes suddenly pale, as for example from shock, the change in skin color is due to a constriction of the arterioles feeding into the capillary network of the skin. Thus blood flow through capillaries is diminished and the skin pales.

It is by virtue of capillaries that nutritive elements (oxygen and nutrients) and waste pass into and out of the tissues, for capillaries are the only vessels thin enough for rapid transfer of small molecular materials. Their single-layer endothelium composition, which varies among the different types of capillaries, and surrounding basement membrane makes capillaries a bit "leakier" than other types of blood vessels. This allows oxygen and other molecules to reach the body's cells with greater ease.

Capillaries connect the **arterial system**—which includes the blood vessels that carry blood away from the heart—to your **venous system**. Your venous system includes the blood vessels that carry blood back to your heart. Additionally, white blood cells from your immune system can use capillaries to reach sites of infection or other inflammatory damage.

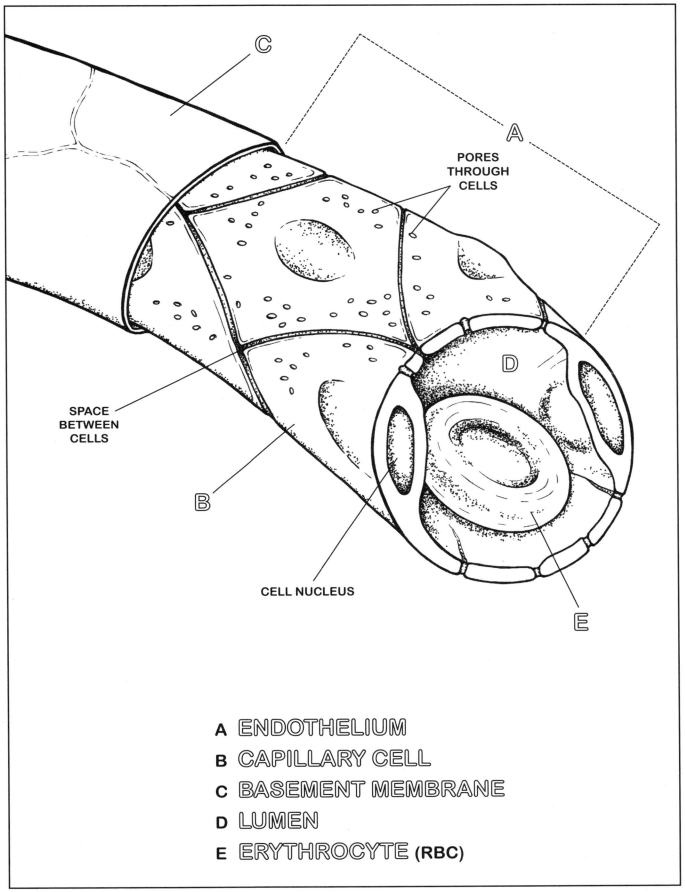

C

A

PORES
THROUGH
CELLS

SPACE
BETWEEN
CELLS

B

D

CELL NUCLEUS

E

A ENDOTHELIUM
B CAPILLARY CELL
C BASEMENT MEMBRANE
D LUMEN
E ERYTHROCYTE (RBC)

THE CARDIOVASCULAR SYSTEM
PORTAL SYSTEMS

In most organs, the blood supply is channeled in by arteries, passes through a capillary plexus, and leaves the organ via veins. In certain regions—e.g., the liver, GI tract, and the pituitary gland—blood passes through two capillary networks before returning to the heart. Such a phenomenon is called a **portal system**. Oxygen and nutritional exchange occurs in the first capillary plexus; the venous blood is then transported into a second capillary network, where the lining cells collect certain materials for processing and secrete other molecules into the venous blood.

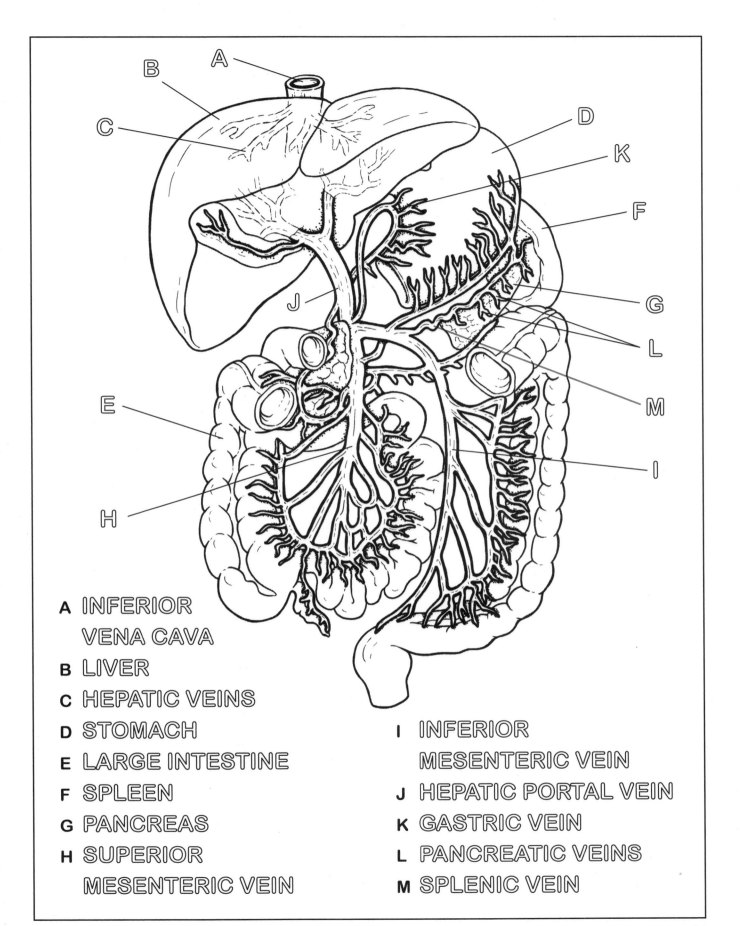

A INFERIOR
VENA CAVA
B LIVER
C HEPATIC VEINS
D STOMACH
E LARGE INTESTINE
F SPLEEN
G PANCREAS
H SUPERIOR
MESENTERIC VEIN

I INFERIOR
MESENTERIC VEIN
J HEPATIC PORTAL VEIN
K GASTRIC VEIN
L PANCREATIC VEINS
M SPLENIC VEIN

THE LYMPHATIC SYSTEM: VESSELS & LYMPH NODES

The molecular concentrations of extracellular fluid (ECF) of the body are in equilibrium with molecules in the plasma (a type of ECF) of the blood. Functioning to help maintain a stable fluid and electrolyte balance among the cells, tissue spaces, and the blood by drawing off excess fluid, fat, and small protein molecules are the **lymphatic vessels**. These vessels make up a closed system of connective tissue-supported endothelial tubes, permeable to water and molecules up to the size of small proteins, and found in most tissues throughout the body, with notable exceptions (e.g., the brain). The smallest of the lymph vessels (capillaries) arise as blind tubes in the connective tissue. These anastomoses form the tributaries of larger lymph vessels.

The movement of lymph is created by local skeletal muscle movement, since no lymphatic "heart" exists, and is made possible by valves in many of the lymph vessels. Lymph vessels, often running in company with neurovascular bundles, are tributaries of larger lymph channels, two of which drain into the **venous system**. The major vessels will be considered regionally. Lymph vessels are but one part of the **lymphatic system**.

Throughout the lymph vessels, usually in specific locations about the body, one finds collections of **lymph nodes**: variable sized, bean-shaped structures that filter the lymph fluid. Lymph nodes are encapsulated structures that are fed lymph fluid by one or more **afferent** vessels and then discharge the lymph fluid into one or more **efferent** vessels at the **hilus** of the node.

The outer third (cortex) of the node is populated by circular **nodules** of lymphatic tissue—masses of lymphocytes set in a dense but delicate nest of reticular fibers. In the **cortex** of each nodule (germinal center) a proliferation of lymphocytes often takes place, particularly when the node is involved in a defensive maneuver. Phagocytes may be found here too.

The more central part of the node, the **medulla**, is characterized by irregular cords of lymphatic tissues through which vast numbers of interlacing lymph channels (**sinuses**) weave their course. These sinuses ultimately all arrive at the **hilus** to form the **efferent lymph vessels**. The sinuses are lined by phagocytic cells which play a starring role in the filtering of foreign bodies (bacteria, etc.) from the circulating lymph.

Thus, lymph nodes not only "strain" the lymph but are important sources of lymphocytes. In times of adversity (infection), lymph nodes can swell three or four times their normal size of approximately ½ inch (12 mm) across. This is a consequence of increasing numbers of lymphocytes within the node. These cells are often distributed peripherally via the lymphatic vessels to fight infection.

Aggregations of lymph nodes occur throughout the body, usually in association with deep or superficial veins. Superficial groups of lymph nodes can be palpated easily—particularly when they are enlarged following a local infection. Such groups include the **cervical** lymph nodes (felt on the side of the neck—they can be rolled underneath the skin), **axillary** lymph nodes (in the "armpit" region), and the **inguinal** lymph nodes (at the groin). These groups are employed by the position as indications of infection in regions drained by each group. Therefore, enlarged cervical nodes are expected during acute or chronic tonsillitis, the axillary nodes may trap metastasizing cells from carcinoma of the breast, and the inguinal nodes may be involved in an abscess of the knee.

LYMPH NODES:

A₁ CERVICAL
A₂ AUXILLIARY
A₃ ABDOMEN
A₄ INGUINAL

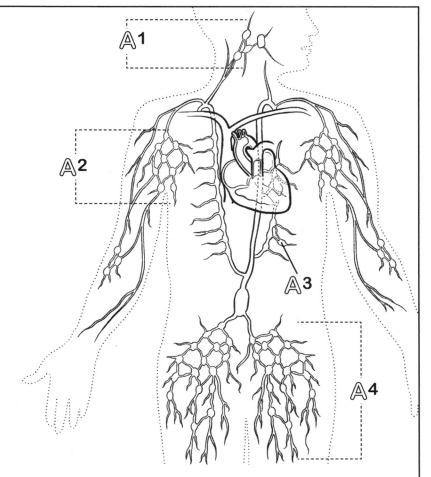

WITHIN LYMPH NODE:

B AFFERENT VESSELS
C EFFERENT VESSELS
D NODULE
E CORTEX
F HILUS
G MEDULLA
H CAPSULE
I BLOOD VESSELS

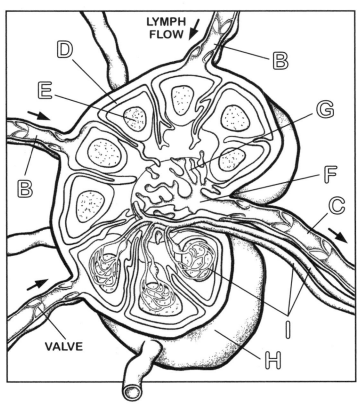

INTRODUCTION TO THE PERIPHERAL NERVOUS SYSTEM (PNS)

The peripheral nervous system (PNS) is that part of the body's nervous tissue organization that deals with:

• Receipt of sensory information by receptors.
• Transmission of the information to the **central nervous system (CNS).**
• Motor commands transmitted from the CNS to the peripheral effectors—glands and skeletal, smooth, and cardiac muscle.

The PNS is composed of:

Spinal nerves: Consisting of sensory nerves, which transmit impulses to the spinal cord portion of the CNS from sensory receptors; and motor nerves, which transmit impulses from the spinal cord to the effector organs. Spinal nerves leave from the spinal column.

Cranial nerves: Consisting of sensory nerves that pass along impulses to the brain portion of the CNS from sensory receptors, and motor nerves that carry impulses from the brain to the effector organs. In the peripheral nervous system, note that cranial nerves leave from the head and neck.

Autonomic nervous system (ANS): A division of the PNS with special responsibility for cardiac muscle and for the smooth muscle and glands of the viscera. The ANS will be discussed separately from the PNS later in this book.

Ganglia: The cell bodies of most sensory neurons of the peripheral nervous system are located in **ganglia** adjacent to but outside of the brain and spinal cord. The cell body, you will remember, is a part of the neuron containing the nucleus and some peripheral cytoplasm. The large masses of these cell bodies, their satellite cells, and axons produce a visible swelling along the dorsal root fibers within the enclosure of the intervertebral foramen.

Spinal and cranial nerves, as well as tracts of the CNS, are composed of axons having one of several possible functions. In this respect, a nerve axon of the PNS is sensory (**afferent**) or motor (**efferent**) in function. If it is associated with viscera, it is designated visceral afferent or visceral efferent. If it is associated with the skin, fascia, or musculoskeletal system, it is designated somatic afferent or somatic efferent.

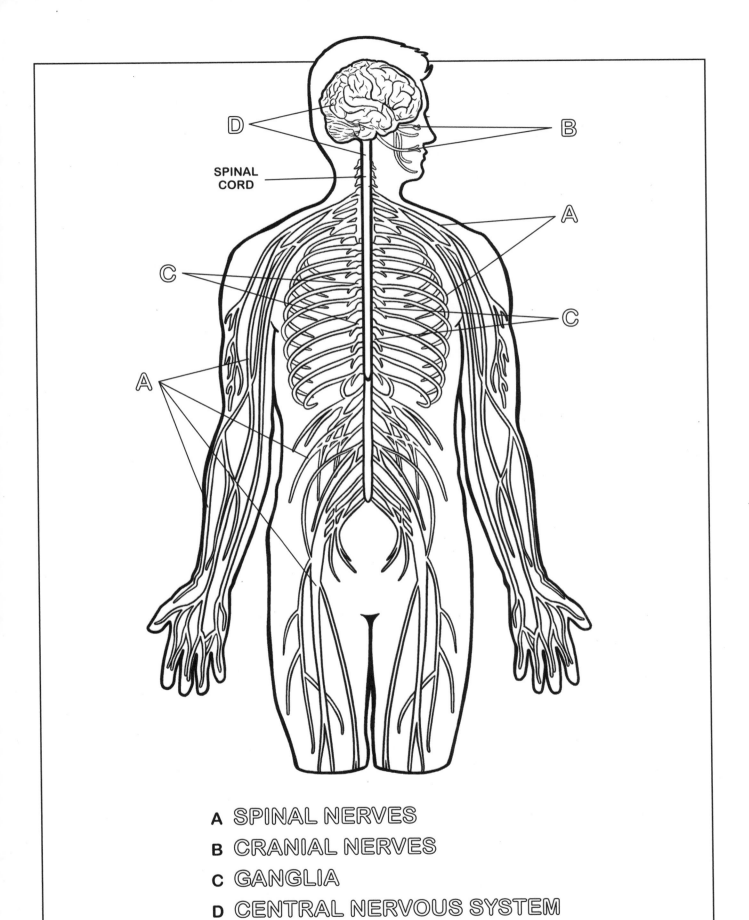

D

B

SPINAL
CORD

A

C

C

A

A SPINAL NERVES
B CRANIAL NERVES
C GANGLIA
D CENTRAL NERVOUS SYSTEM

THE PERIPHERAL NERVOUS SYSTEM
SPINAL NERVES

A **spinal nerve** is composed of variably sized, functionally mixed axons of the PNS. Formed from two spinal roots emerging from the spinal cord, spinal nerves are sheltered by the bony vertebral column. One root, the **sensory root**, leaves the cord on its dorsal surface, and so it is called the **dorsal root**. The **motor root** leaves the cord on its ventral surface and is therefore called the **ventral root**. The two roots come together about 1 inch lateral to the cord to form a **spinal nerve**.

The spinal nerve then divides into an **anterior ramus** and a **posterior ramus** (pl., rami, branches). The posterior ramus, usually the smaller of the two, serves the musculoskeletal regions of the back, while the anterior ramus serves the anterolateral body wall and the limbs.

There are 32 pairs of spinal nerves exiting the spinal cord from skull to coccyx. The nerves are divided into regions similar to the regional breakdown of vertebrae:

Cervical: Eight spinal nerves
Thoracic: Twelve spinal nerves
Lumbar: Five spinal nerves
Sacral: Five spinal nerves
Coccygeal: Two spinal nerves

The precise arrangement and distribution of the nerves differ from region to region; however, the thoracic spinal nerve is usually employed as a classical spinal nerve.

In general, the spinal nerves form networks called **plexuses** before distributing out to peripheral organs. The major plexuses of peripheral nerves are formed by anterior rami.

Each spinal nerve is distributed to a specific region of the body. The ventral root contributes efferent (motor) fibers, both somatic and visceral, to the muscles, glands, and the viscera of that region. The dorsal root is composed of afferent sensory fibers coming in from pain, temperature, pressure, touch, and the proprioception receptors of that region. The body surface region served by one dorsal root of a spinal nerve is called a **dermatome**. An appreciation of dermatomes is particularly important to the physician. A nerve dysfunction may be manifested by loss of some or all sensation in one or more dermatomes. Furthermore, pain in a dermatome may actually be caused by disorder not at the body surface but in one of the viscera. This is called **referred pain**. The dermatome affected is the one with sensory fibers entering the cord to the same dorsal root as the afferent fibers from the injured organ.

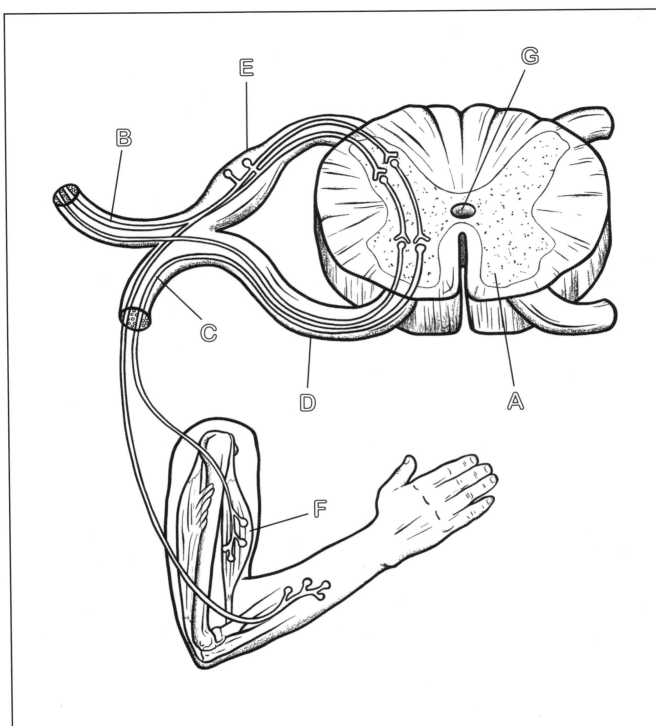

A SPINAL NERVE
B DORSAL ROOT
C VENTRAL ROOT
D ANTERIOR RAMUS

E POSTERIOR RAMUS
F MUSCLE
G CENTRAL CANAL

THE PERIPHERAL NERVOUS SYSTEM
CRANIAL NERVES

Structurally and functionally, **cranial nerves** are similar to spinal nerves. They are just related to different parts of the CNS. The twelve cranial nerves, with two exceptions, serve only the head region. Unlike spinal nerves, each cranial nerve is named, usually by function or destination.

The twelve cranial nerves are shown opposite.

Many of the cranial nerve sensory ganglia are comparable to spinal ganglia, i.e., they are unipolar. However, the ganglia of the olfactory, optic, and vestibulocochlear (I, II, VII) nerves are bipolar cells located near the receptor organs. Motor ganglia, associated with the autonomic nervous system, will be discussed later in this book.

CRANIAL NERVES

I. OLFACTORY

II. OPTIC

III. OCULOMOTOR

IV. TROCHLEAR

V. TRIGEMINAL

VI. ABDUCENS

VII. FACIAL

VIII. VESTIBULOCOCHLEAR

IX. GLOSSOPHARYNGEAL

X. VAGUS

XI. ACCESSORY

XII. HYPOGLOSSAL

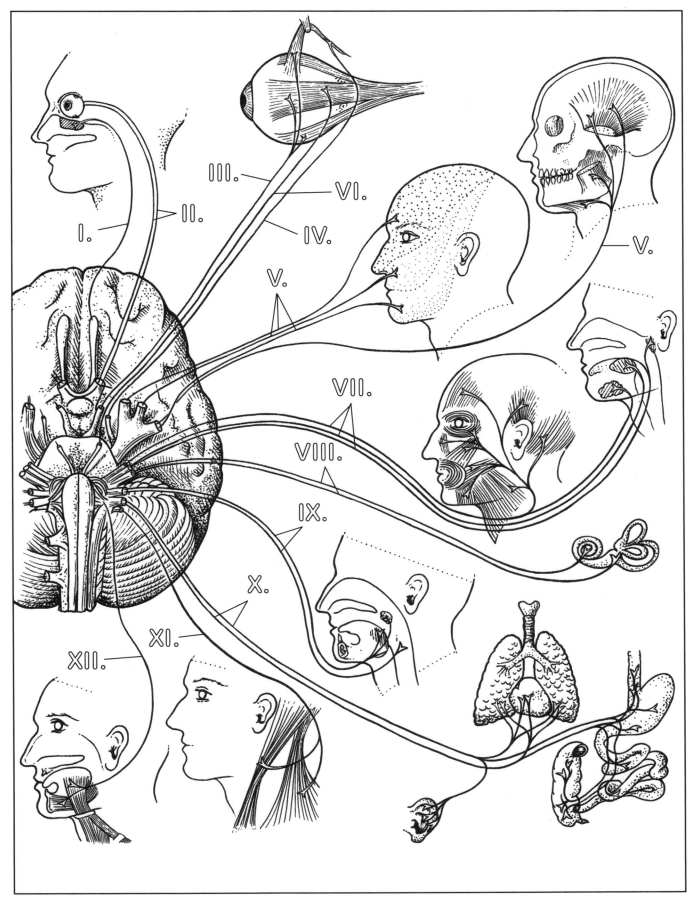

I.

II.

III.

IV.

V.

V.

VI.

VII.

VIII.

IX.

X.

XI.

XII.

INTRODUCTION TO THE AUTONOMIC NERVOUS SYSTEM (ANS)

The **autonomic nervous system** is responsible for distributing motor impulses to viscera, and consists of visceral efferent motion fibers and their cell bodies. You are now familiar with the concept that the PNS is primarily a voluntary activated system. The autonomic nervous system is predominantly an involuntary system capable, under certain conditions, of voluntary activation as well.

As your study of body structure progresses, you will find that most of the viscera of the body have the capability to move, secrete, or both. All of the viscera of the alimentary canal (stomach, intestines, etc.) have the capacity to churn their contents mechanically (and to secrete enzymes as well) in order to assist in the breakdown of these contents. They do this through the smooth muscle and glands within their walls. You may know that this digestive process takes place without any voluntary action on your part.

Likewise, the viscera of the respiratory system (lungs, etc.), cardiovascular system (heart, blood vessels), urinary system (kidneys, bladder, etc.), and the reproductive system (glands and ducts), as well as the various glands throughout the body, also have the capacity to move, secrete, or both—and do for a lifetime—without any conscious effort on your part. Such visceral activity is the responsibility of the autonomic nervous system.

When quiet relaxation is the order of the day, the autonomic nervous system acts to maintain the internal environment accordingly: slower respiration rates, slower heart rate, increasing the digestive processes, and so on. Yet when there is a threat to one's survival, the autonomic nervous system takes appropriate action: increasing blood flow to the skeletal muscles involved in "flight," increasing respiration rate and other actions designed to allow you to run faster than you thought you could or take other evasive actions.

The maintenance of a stable internal environment relative to outside environmental conditions is called homeostasis, and that, in a word, is the function of the ANS.

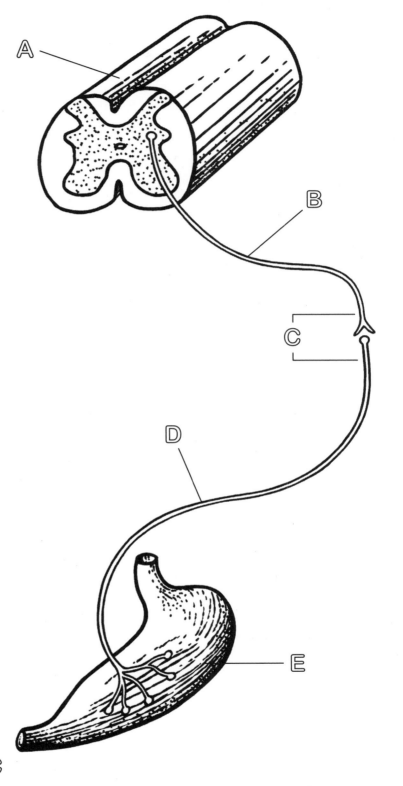

A SPINAL CORD
B PREGANGLIONIC
   NEURON
C GANGLION
D POSTGANGLIONIC
   NEURON
E EFFECTOR ORGAN

REFLEXES

In the functional sense, a **reflex** is an involuntary motor response to impulses generated by sensory receptors. We demonstrate various reflexes constantly throughout the day and night. Reflexes are mechanisms by which our various body parts and organs can function without our conscious awareness. There are certain basic functions of the body that must go on, for example, when we are asleep: visceral movements during digestion, with changes in heart and respiratory rates, musculoskeletal movement in response to itching, irritation, or pain, etc. These actions are accomplished by means of reflexes. In a reflexive manner we move out of the way of a speeding object or we set our bodies to pick up heavy weights, all without thinking about it.

By definition, a reflex requires two structural components: an **afferent** limb and an **efferent** limb. The afferent limb starts with a sensory receptor and ends in the CNS via an axon. The efferent limb starts in the CNS—directly or indirectly connected to the afferent limb—and ends at the effector organ via an axon. This CNS, then, is the key to all reflexes.

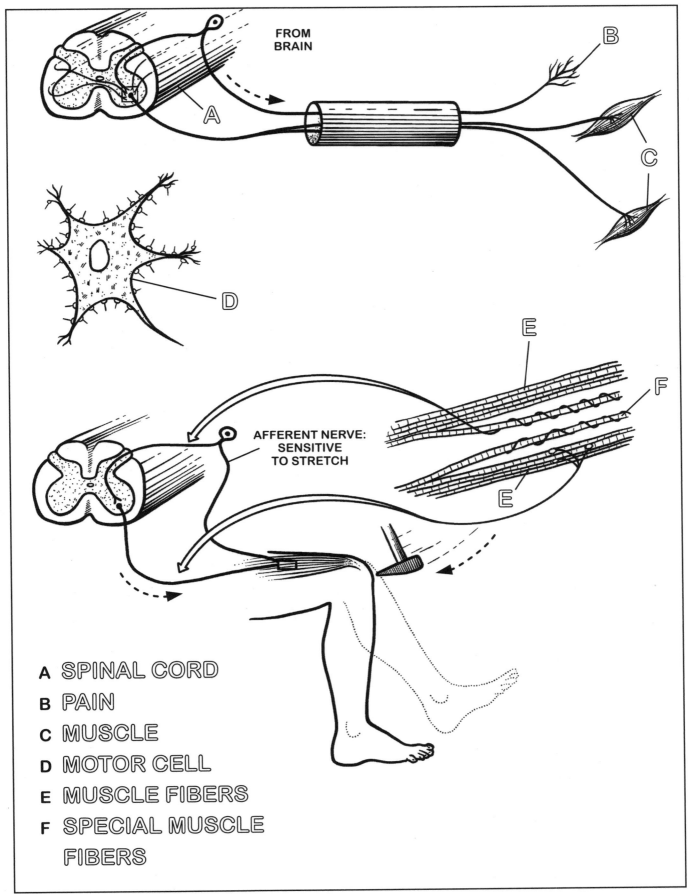

FROM BRAIN

AFFERENT NERVE: SENSITIVE TO STRETCH

A SPINAL CORD
B PAIN
C MUSCLE
D MOTOR CELL
E MUSCLE FIBERS
F SPECIAL MUSCLE FIBERS

SKIN:
EPIDERMIS

There is no magician's mantle to compare with the skin in its diverse roles of waterproof overcoat, sun shade, suit of armor, and refrigerator, sensitive to the touch of a feather, to temperature, and to pain, withstanding the wear and tear of three scores and ten, and executing its own running "repairs."

The truth here is as profound as the description is elegant. Further introduction would be redundant.

The skin of the body (integument) is an external surface lining, continuous with the internal surface membrane of those body cavities open to the exterior. This smooth, uninterrupted junction of the external skin of the lip with its internal surface in the mouth is an excellent example of an external-internal continuity which can be seen and felt on yourself. Note how dry the external portion of the lip is and how moist the internal portion is. The latter surface is a mucous membrane which continues caudally to line the entire gastrointestinal tract and is continuous with the skin of the anus as well.

The integument is arranged into two well-defined layers:
• **Epidermis**: the outer, stratified epithelial layer.
• **Dermis**: the connective tissue layer underlying the epidermis.

EPIDERMIS
The epidermis is simply stratified squamous epithelium. Slight differences in the character of the epithelium may be seen in any cross section, thus creating strata or layers.

The **stratum basale** is the basal (deepest) layer of cells of the epidermis and is the germinating layer— columnar cells that divide constantly and whose progeny push up toward the free surface to replace the cells that have died and been sloughed off. These cells are closest to the underlying dermis and the capillary beds therein, thus their sustenance is assured.

The upper layers of dead cells are **stratum corneum** and they make up the majority of the epidermis in thick skin. Nails, at the tips of fingers and toes on the posterior surface of the body, are composed of modified cells of stratum corneum. They are diminutive forms of claws and hooks of man's distant ancestors.

The epidermis is entirely without blood vessels, penetrated only by the free endings of pain receptors and the coiled ducts of sweat glands.

DERMIS
The dermis is bound to deeper structures (e.g., periosteum or deep fascia) by a variable layer of subcutaneous tissue, the superficial fascia. This layer may be thick or thin; for example, contrast on yourself the depth or amount of fascia between the skin and periosteum on your shin, and the amount of fascia between skin and underlying muscle at your waist just above the buttocks. We will investigate the dermal layer of the skin on the following page.

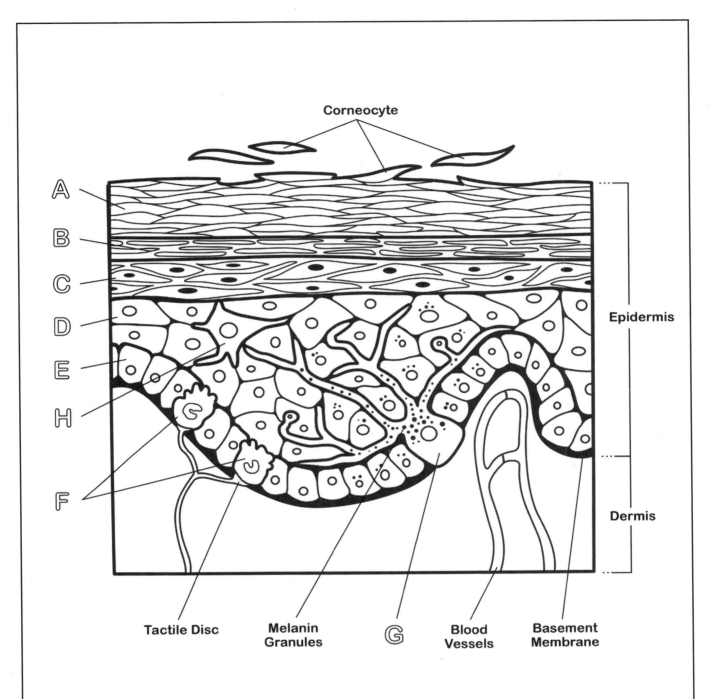

Corneocyte

A

B

C

D

E

H

F

Epidermis

Dermis

Tactile Disc

Melanin
Granules

G

Blood
Vessels

Basement
Membrane

A STRATUM CORNEUM

B STRATUM LUCIDUM

C STRATUM GRANULOSUM

D STRATUM SPINOSUM

E STRATUM BASALE

F MERKEL CELL

G MELANOCYTE

H LANGERHANS CELL

DERMIS & APPENDAGES OF THE SKIN: HAIR & GLANDS

DERMIS

The dermis is the connective layer underlying the epidermis; it is highly vascular with **papillae** projecting into the epidermis. These papillae are characterized by capillary nets and certain sensory receptors. Arising in the dermis are hair follicles and coiled sweat glands, whose shafts and ducts, respectively, pass upward through the epidermis. In the dermis, associated with each hair shaft, is an oil or sebaceous gland and a small bundle of smooth muscle (**arrector pili**). The dermis is replete with sensory nerve fibers conducting sensory impulses from intradermal receptors and with ANS (sympathetic) motor fibers innervating the arrector pili muscles. The dermis is continuous below with the superficial fascia.

HAIR

Most of the human body is covered with **hair**, even when not immediately apparent. It would seem that hair on people is largely **vestigial** (having become functionless through evolution), but in other mammals, hair plays an important role in protection and heat insulation. In people, there may be significance in the fact that touch receptors are immediately associated with the hair shaft.

A hair is an elongated shaft of dried, keratinized cells and a root arising from its surrounding **follicle**. The follicle, formed from an invaginated bud of surface epithelium in the embryo, is the germinating component of the hair. The hair grows lengthwise by mitosis occurring at the root. Hair exits onto the skin surface through the "passageway" created earlier by the invaginating epidermal bud. As long as the follicle is not injured or removed, the hair shaft will grow, regardless of how many times it is cut. Actually, a hair will stop growing only when its follicle has ceased mitotic activity, a phenomenon under genetic control. Periodically, the major portion of the follicle dies and vanishes. Some of the remaining cells then form a new follicle and a new hair is generated, forcing the old (club) hair out. The development, growth, and recession of certain hairs likely comes under the influence of sex hormones (testosterone in the male and estrogen in the female). When these hormones appear in significant concentrations in the blood at puberty, development of the coarse hairs of the axillary, pubic, and (in the case of the male) other organs commences. Loss of scalp follicles or diminution in their activity in the male, causing baldness (alopecia), may also be related to the onset of puberty.

GLANDS

In general the skin contains two kinds of glands: **sweat glands**, which are simple, coiled glands located usually in the deep dermis; and **sebaceous glands**, which are simple, sac-shaped exocrine glands of the dermis whose ducts open into the distal sheets of hair follicles.

The coiled portions of sweat glands are secretory, composed of pyramid-shaped columnar cells and closely related myoepithelial cells. The latter cells are innervated by fibers of the sympathetic division of the autonomic nervous system. When stimulated, the processes of these cells, oriented around the secretory cells, "squeeze" the secretions from the cells into the ducts. This salty, watery fluid is transported up through the corkscrew duct of each gland and deposited onto the skin surface. The general secretory product of a sweat gland is perspiration. The secretory activity of sweat glands, as a way of releasing heat, plays an important role in the body's temperature regulating mechanism.

Sebaceous glands are absent in regions without hair. Each sac, termed an **alveolus**, consists of a mass of epithelial cells; the lining cells of this alveolus divide frequently and their progeny are pushed into the center. These cells ultimately break down and become the oily secretory product (sebum) of the alveolus. Excretion of sebum onto the skin surface occurs through the action of the **arrector pili muscles** (which are smooth muscles) attached to the sheath of each hair. Note that there are specializations of these glands in the eyelids, the **tarsal glands**. The purpose of sebum is to protect the skin from dehydration, but in the eyelids, the sebaceous gland secretes a special type of sebum into tears.

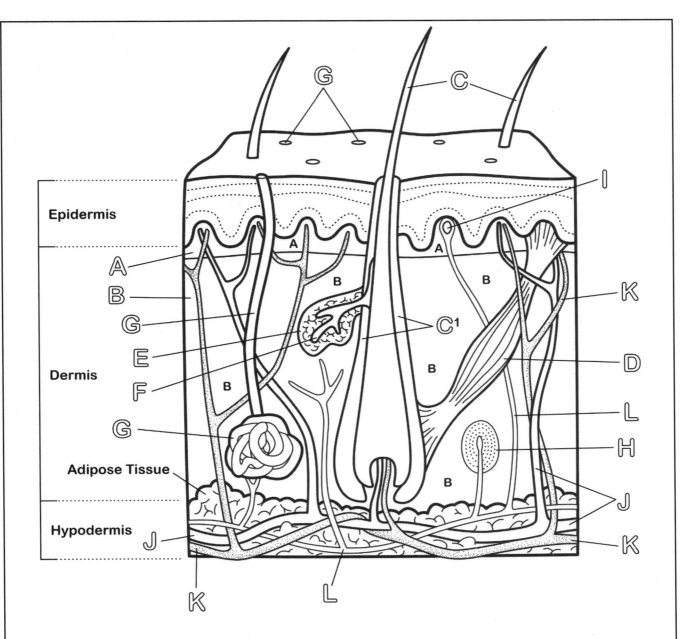

Epidermis

Dermis

Adipose Tissue

Hypodermis

A PAPILLARY LAYER

B RETICULAR LAYER

C HAIR SHAFT

  C¹ FOLLICLE

D ARRECTOR PILI MUSCLE

E SEBACEOUS GLAND

F SEBUM

G SWEAT GLAND

H PACINIAN CORPUSCLE

I MEISSNER'S CORPUSCLE

J ARTERY

K VEIN

L NERVE

SECTION FOUR:
THE UPPER LIMB

The upper limbs are truly a marvelous set of structures. Consider some of the tasks we can accomplish with them: We can climb trees using our upper limbs as stabilizers and as hoisters as we go from one level to another. We can throw a baseball with amazing speed and accuracy, and loft it 100 yards with the end of a bat. The knife-thrower at the circus depends (as does the throwee) upon the delicate neuromuscular coordination of the upper limb. In golf we work the shoulder and wrist joints with powerful muscular movements to effect a 200- or 300-yard drive, and ask for subtle nuances of upper limb movement for that 6-inch putt. And who can tell a story without appropriate gestures of the hands? What is a better use of the upper limbs than to wrap our arms around a loved one to show how much we care?

The upper limb is a functional unit of the upper body. It consists of three sections, the upper arm, forearm, and hand. It extends from the shoulder joint to the fingers and contains thirty bones. It also consists of many nerves, blood vessels, and muscles. The nerves of the arm are supplied by one of the two major nerve plexuses of the human body, the brachial plexus.

The anatomical basis for these and like things is in the next section.

OSTEOLOGY: BONES OF THE UPPER LIMB

You remember that the bony framework of the body is called the skeleton. That part of the skeleton along the principal axis of a body is the axial skeleton, while the bones of the limbs and their bony girdles comprise the appendicular skeleton.

The **upper limb**, seen here, is attached to the axial skeleton by the **pectoral girdle**, composed of two **clavicles** anteriorly and two **scapulae** posteriorly.

The **clavicle** articulates with the **scapula** laterally at the **acromioclavicular joint**. The arm bone, or **humerus**, articulates with **glenoid fossa**, a hollow or depressed area of the scapula. The bones of the forearm, the **radius** and **ulna**, articulate above the humerus and below the wrist bones.

The wrist, or **carpus**, is composed of eight small bones of a variety of assorted shapes, and these bones are named according to their shapes.

The distal row of the carpal bones articulates with the five **metacarpal** bones that make up the skeleton of the palm and dorsum of the hand.

Each metacarpal articulates with a proximal **phalanx** of a digit. There are three phalanges in each digit of the hand, with the exception of the thumb, which has two phalanges.

CLINICAL CONSIDERATIONS OF THE UPPER LIMB: FRACTURES

A **fracture** is simply a break in a periosteum-lined bone. Fractures are caused by trauma or disease (cancer, infection, or metabolic disorders).

Complications from fractures are generated when surrounding soft tissues or neurovascular structures are penetrated. Fractures that break the skin are called **compound** or **open** fractures, and **simple** fractures are when the skin is not broken. Some fractures involve complete breaks in the bone (a **complete** fracture); some have breaks in the bone with the bone still in one piece (an **incomplete** fracture).

Within each of these categories, there are variations in the appearance of the fracture. Some are **impacted** (telescopeal), while others are **comminuted** (broken into several fragments), etc.

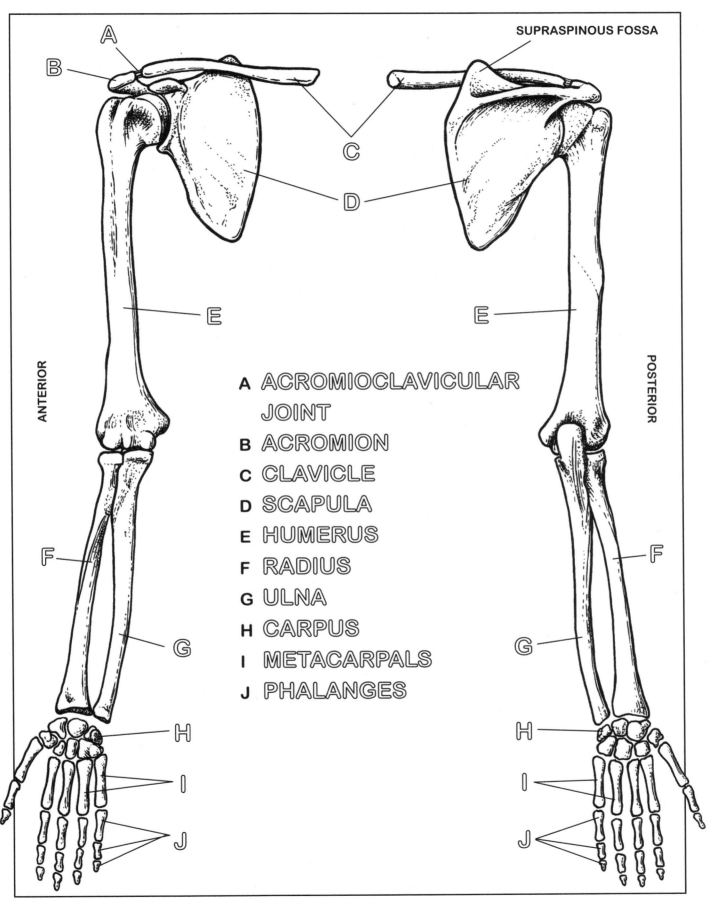

SUPRASPINOUS FOSSA

ANTERIOR

POSTERIOR

A ACROMIOCLAVICULAR
  JOINT
B ACROMION
C CLAVICLE
D SCAPULA
E HUMERUS
F RADIUS
G ULNA
H CARPUS
I METACARPALS
J PHALANGES

CLAVICLE & SCAPULA

CLAVICLE (COLLARBONE)

The clavicle is the anteriormost bone of the pectoral girdle and is often referred to as the *strut* of the upper limb, for it forces the scapula backward and laterally away from the chest wall.

The region around the clavicle-scapula articulation is the shoulder. Medially the clavicle articulates with the **sternum**, or **sternoclavicular joint**. Note the relationship of the clavicle to the **scapula** and the **humerus**. One can understand why, in falls on the hands or violent blows against the shoulder, this strut of the upper limb may snap—and frequently does.

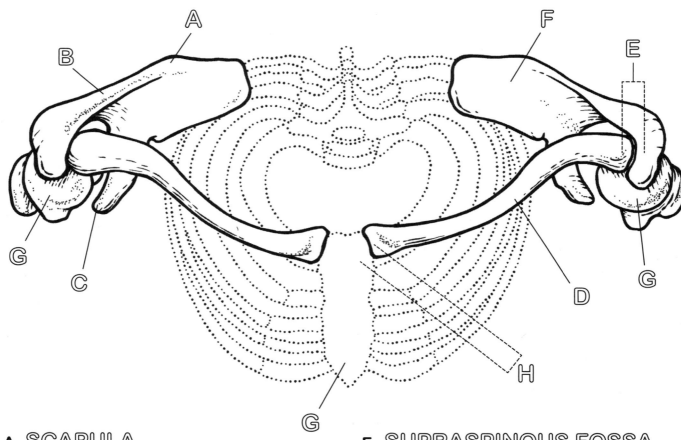

A SCAPULA	**F** SUPRASPINOUS FOSSA
B SPINE OF SCAPULA	**G** HUMERUS
C CORACOID PROCESS	**H** STERNOCLAVICULAR JOINT
D CLAVICLE	**I** ACRIMION
E ACROMIOCLAVICULAR JOINT	**J** GLENOID FOSSA

SCAPULA (SHOULDER BLADE)

The **scapula**, the posteriormost bone of the shoulder girdle, is a flat bone with various surfaces and borders, offering points of attachment for many muscles—seventeen in all! This brings to light a rather important concept for the pectoral girdle: The scapula has no direct bony connection with the axial skeleton—only indirectly, via the clavicle. Yet the humerus articulates with the scapula, which seems to imply that the integrity of the whole upper limb is dependent upon this flimsy sternoclavicular joint. Obviously, as anyone who has done a pull-up will testify, it is not. For if the sternoclavicular joint alone secured the upper limb, the pectoral girdle would literally slip off the thorax with every pull-up.

Some other tissue must be involved, and there is: muscle tissue. It is largely because the muscles securing the scapula to the axial skeleton are capable of a wide range of movement, as we shall learn. This is important, for if the scapula were not capable of rotation, elevation, or depression, freedom of movement of the upper limb would be severely restricted.

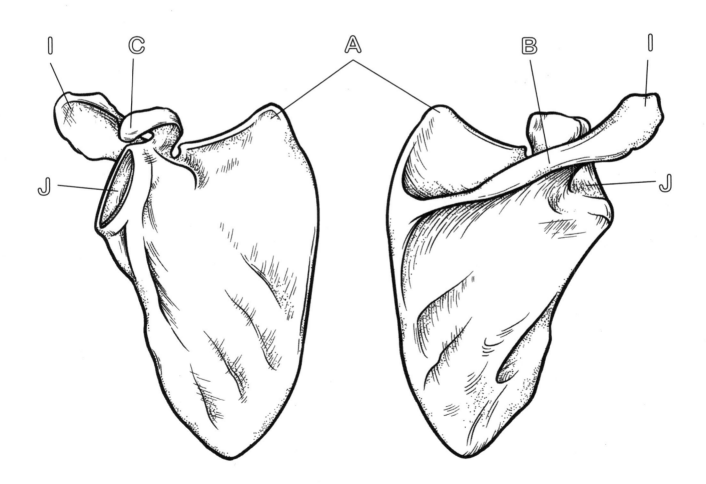

OSTEOLOGY OF THE UPPER LIMB
HUMERUS

HUMERUS

The humerus is the bone of the arm and is easily identified by its "bald" head above and the pulley-shaped articulating **condyle** (a rounded protuberance at the end of some bones) referred to as the **trochlea** (a structure resembling or acting like a pulley) distally. The humerus articulates with the glenoid fossa of the scapula superiorly and with the two bones of the forearm below, the ulna and radius.

The humerus can be palpated in various places throughout the arm; about two fingersbreadth below the acromion anteriorly, one can feel a small projection of bone. This is the lesser tubercle, which receives the attachment of an important muscle. The **greater tubercle**, where three muscles attach, can be felt at the tips of the shoulder just below the acromion laterally. The shaft of the bone can be felt throughout from side to side, including the prominent **deltoid tuberosity** (a bump about halfway down the lateral aspect of the arm) where the deltoid (shoulder muscle) inserts. At the elbow, the **medial epicondyle** is very prominent and feeling its posterior aspect may result in tingling sensations in the fourth and fifth fingers, which means that you have pressed the **ulnar nerve**.

The humerus is subject to fracture and displacement following falls and violent motions, such as throwing a ball. The displacement of the two broken ends occurs because of muscular pull. Dislocation of the shoulder joint can also occur. The head of the humerus usually moves inferiorly and then forward or backward depending on the force.

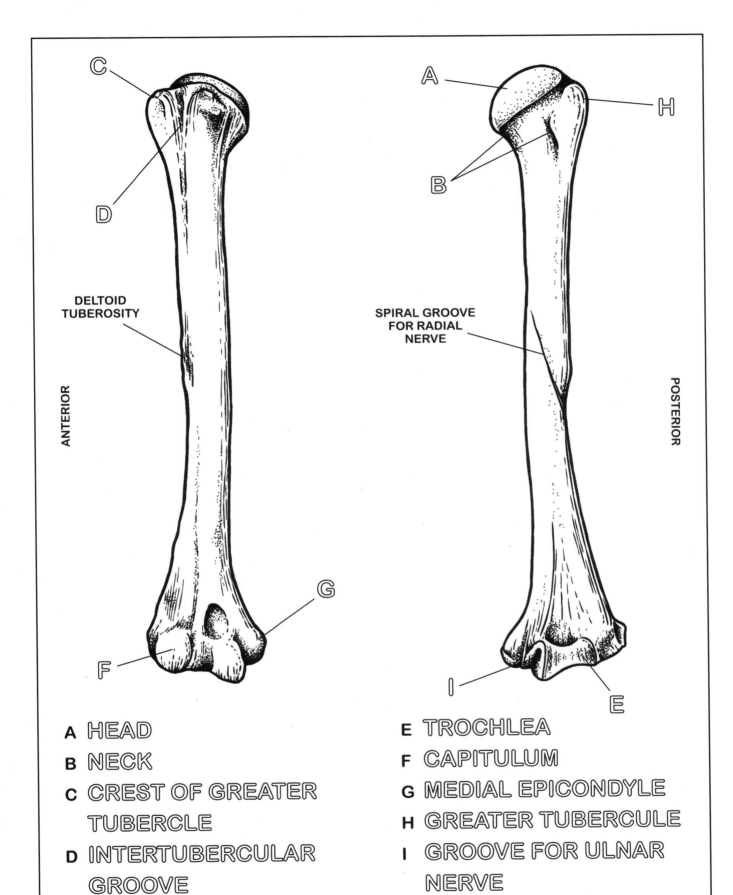

DELTOID
TUBEROSITY

ANTERIOR

SPIRAL GROOVE
FOR RADIAL
NERVE

POSTERIOR

A HEAD

B NECK

C CREST OF GREATER
TUBERCLE

D INTERTUBERCULAR
GROOVE

E TROCHLEA

F CAPITULUM

G MEDIAL EPICONDYLE

H GREATER TUBERCULE

I GROOVE FOR ULNAR
NERVE

OSTEOLOGY OF THE UPPER LIMB
ULNA & RADIUS

ULNA

The **ulna** is one of the two bones of the forearm. In the anatomical position, it occupies the medial (little finger) side of the forearm. The ulna is large above, articulating with the trochlea of the humerus, and slender and small below, where it articulates with the radius directly and the wrist indirectly via articular disc.

The ulna is further characterized by an unusual C-shaped process superiorly: the **olecranon**. The olecranon can best be palpated when the elbow joint is flexed. Note the olecranon does not move when you rotate the forearm from **pronation** to **supination** and back. Try it! Thus the humeroulnar joint is a true hinge joint and the ulna does not rotate.

The posterior aspect of the ulna can be felt throughout the forearm, almost down to the wrist. Here it ends as a rounded projection of bone, easily felt on the posterior aspect of the forearm just above the wrist, medially. This is the **styloid process** when the forearm is supinated—it is the head of the ulna when the forearm is pronated. With your forearm pronated, place a finger on the head of the ulna (a significant interior bump at the little finger side of the wrist). Now, rotate laterally. The projection largely disappears, and what you feel is the styloid process of the ulna. What's happening? It is the fact that the ulna does not rotate, but movement of skin and tendons uncover the head of the ulna as the forearm is pronated.

RADUIS

The **radius** is that bone of the forearm occupying the lateral or thumb side. A rounded or tabletop-shaped head at the proximal end is its distinguishing characteristic. In contrast to the ulna, the radius thickens from above to below to form a stout distal extremity which articulates with two bones of the wrist, as well as the ulna. Proximally the radius articulates with the capitulum (head-like process) of the humerus, as well as the radial notch of the ulna. The radius is largely covered with muscle except at its distal extremity, where its prominent lateral border (remember the anatomical position) can be felt down to the lateral border of its styloid process.

Note that when you rotate the forearm, the radius follows the thumb. So extend the elbow joint with the limb supinated (anatomical position) and note that the ulna is definitely medial to the radius throughout the forearm. Now, keep your fingers on the olecranon of the ulna and slowly pronate (medially rotate) the hand (without rotating your arm), keeping your eye on the distal, lateral border of the radius. You can see that the radius crosses over the distal two thirds of the ulna in the act of pronation. The ulna, as you can feel, does not rotate. Is the radiohumeral joint adapted for rotation? It sure is, for the tabletop head of the radius pivots under the capitulum of the humerus. This is a pivot type of synovial joint.

The radius is a common site of fracture, usually caused by falling on outstretched hands.

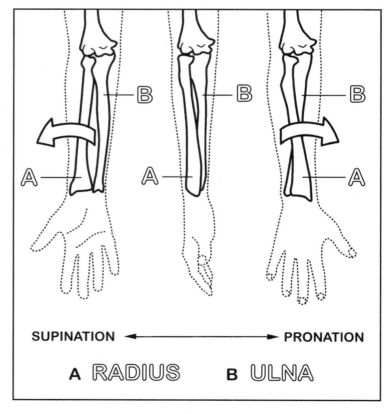

SUPINATION ⟵⟶ **PRONATION**

A RADIUS   **B** ULNA

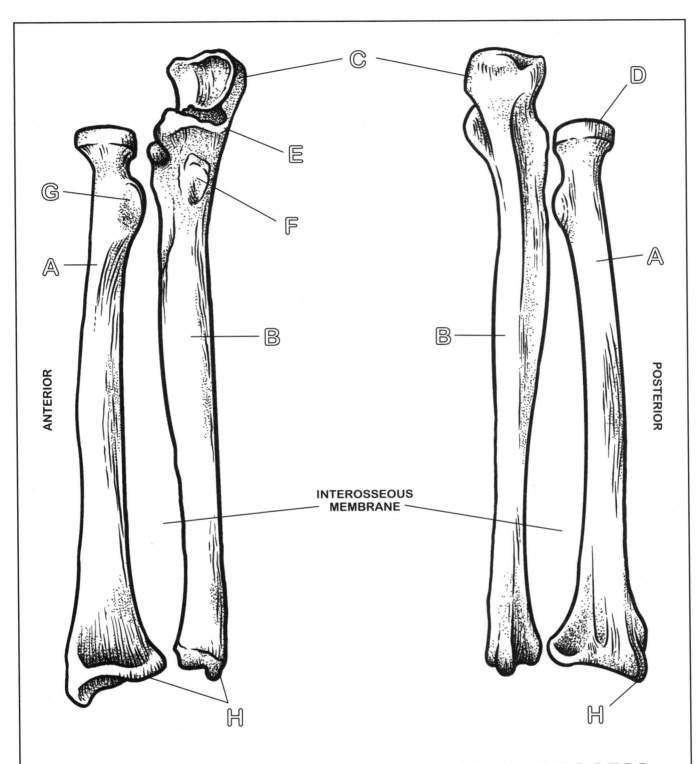

ANTERIOR

POSTERIOR

INTEROSSEOUS
MEMBRANE

A RADIUS
B ULNA
C OLECRANON PROCESS
D HEAD

E CORONOID PROCESS
F TUBEROSITY OF ULNA
G RADICAL TUBEROSITY
H STYLOID PROCESS

OSTEOLOGY OF THE UPPER LIMB
CARPUS, METACARPALS & PHALANGES

CARPUS

The **carpus**, or wrist, is the proximal part of the hand and is composed of eight bones. The carpal bones are arranged in two rows of four each. Each carpal bone has its own distinguishing characteristic and is named accordingly:

PROXIMAL ROW:

scaphoid	**lunate**	**triquetrum**	**pisiform**
(bone-shaped)	(moon-shaped)	(three-sided)	(pea-shaped)

DISTAL ROW:

trapezium	**trapezoid**	**capitate**	**hamate**
(little table)	(table-shaped)	(head-like)	(hook)

There is little value, if any, in studying the individual wrist bones. They should be studied collectively as an entity. The scaphoid and lunate articulate with the radius, while the triquetrum articulates with an articular disc distal to the ulna. Of all the bones of the wrist, the scaphoid seems to be the most subject to fracture.

METACARPALS

Metacarpals are the long bones of the palm of the hand. There are five of them—one for each finger. Their bases articulate with the distal carpal row proximally (carpometacarpal joints) and their heads with each proximal phalanx distally (metacarpo-phalangeal joints). These bones are more easily felt on the posterior surface. Their heads are felt distally as knuckles in a clenched fist. Can you palpate each metacarpal?

PHALANGES

Phalanges are the bones of the **digits**, or fingers. Each bone is a **phalanx**. Flex your finger slightly and look at your hands. You can see that there are three phalanges in each digit with the exception of the thumb, which has only two. By experimenting, you will find that the interphalangeal joints are of the hinge variety.

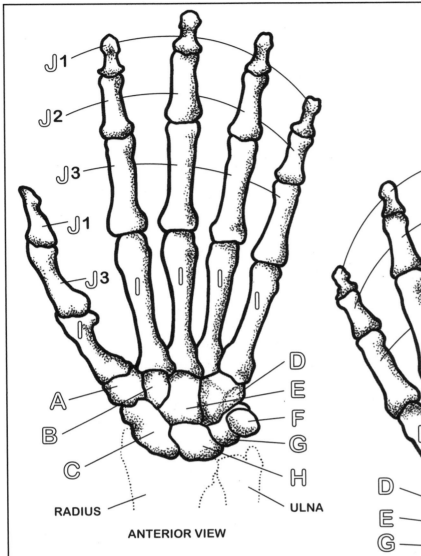

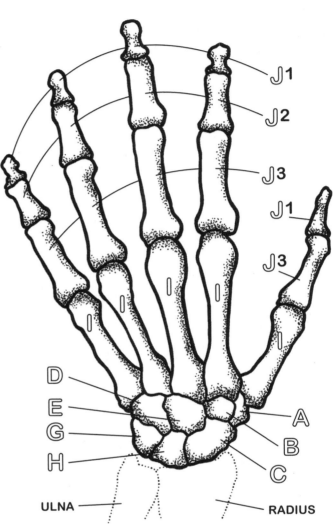

ANTERIOR VIEW

POSTERIOR VIEW

CARPAL BONES

A TRAPEZIUM

B TRAPEZOID

C SCAPHOID

D HAMATE

E CAPPITATE

F PISIFORM

G TRIQUETRUM

H LUNATE

BONES OF THE HAND

I METACARPALS

J PHALANGES

  J1 DISTAL

  J2 MIDDLE

  J3 PROXIMAL

MYOLOGY OF THE UPPER LIMB: MUSCLES "MOORING" SCAPULA TO AXIAL SKELETON

The scapula, supported by muscles from the axial skeleton, is capable of a number of movements, usually associated with simultaneous movements of the arm. You can demonstrate to yourself some of the movements of the scapula. Sit or stand erect, shoulders back. Now move your shoulders forward and medially (as if trying to keep warm) and at the same time place your fingers on your clavicle. Notice how the clavicle's lateral extremity is pushed forward and upward. Since the scapula articulates with the lateral end of the clavicle, it follows that if the clavicle is moved upward and forward, the scapula must have moved with it. Now continue moving your shoulders forward and backward while observing the tip of your shoulder (acromion of the scapula). The muscles that caused the scapular movements are some of the "mooring" muscles of the scapulae. They all arise from the axial skeleton and they all insert on the scapula. To securely moor the scapula, it would seem reasonable to assume that the muscles inserting on the scapula must arise laterally, medially, superiorly, anteriorly, and inferiorly to it, and they do.

Of the five muscles supporting the scapula, the largest is a superficial, trapezoid-shaped muscle that provides the slope of the shoulder from the neck and gives both the back and the upper back their form. Obviously then, this muscle arises superiorly and medially to the scapula. It also arises inferiorly to its scapular attachment. This is the **trapezius** muscle.

Immediately deep to trapezius between the two scapulae there is a rhomboid-shaped muscle on each side of the vertebral column attaching to the vertebral border of the scapula. These muscles reinforce the action of the middle fibers of the trapezius by retracting the scapula. These muscles are called **rhomboids**. They cannot be felt.

Also hidden under the cover of the trapezius immediately superior to the rhomboids is a strap-like muscle inserting at the superior angle of the scapula and arising from the upper vertebrae of the neck. It serves as a superior mooring muscle of the scapula as well as acting to elevate it—hence its name, **levator scapulae**.

The lateral mooring muscle of the scapula is one that arises from the lateral aspect of ribs one to eight and in so doing, appears saw-toothed or serrated on the body surface. Being the anteriormost of two such serrated muscles, you might suspect its name: **serratus anterior.** Serratus anterior can frequently be felt contracting one and a half handbreadths below the axilla when the arm/shoulder is abducted against resistance, for the muscle is principally a *protractor* of the scapula (pulls the scapula laterally about the rib cage).

The anterior mooring muscle of the scapula arises from the anterior chest wall (pectoral region) and is the smaller and deeper of two such muscles in the region. From this you might guess its name: **pectoralis minor.** This muscle inserts on the coracoid process of the scapula, but it cannot be palpated because it lies under the cover of its larger fellow, **pectoralis major**. Pectoralis minor functions in depression and protraction of the scapula.

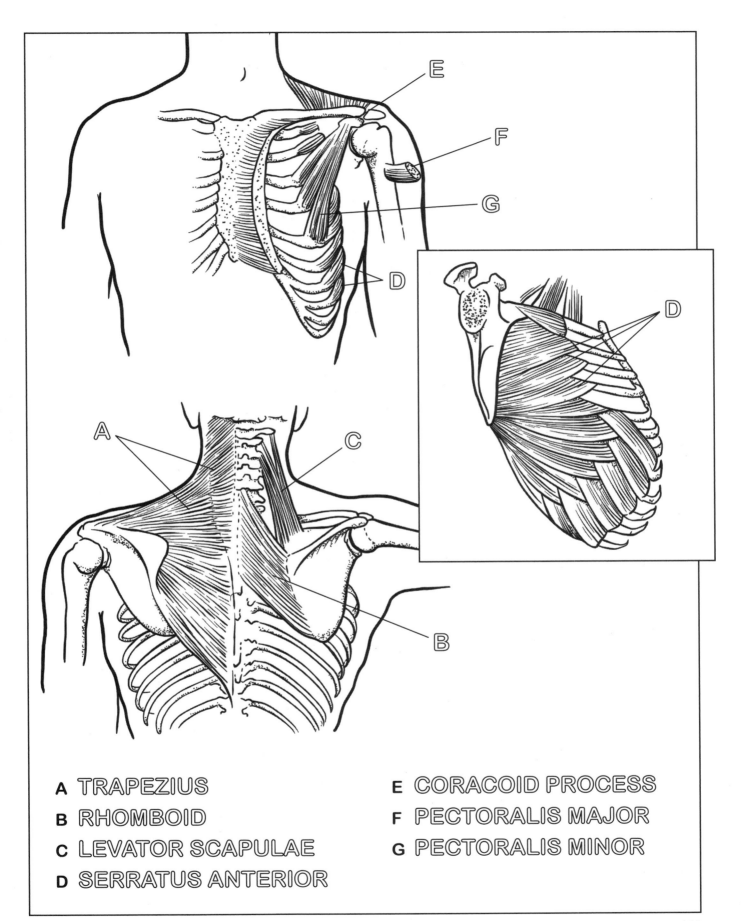

A TRAPEZIUS

B RHOMBOID

C LEVATOR SCAPULAE

D SERRATUS ANTERIOR

E CORACOID PROCESS

F PECTORALIS MAJOR

G PECTORALIS MINOR

MUSCLES SECURING HEAD OF HUMERUS TO SCAPULA

At this point, you should acquaint yourself with the shoulder, or **glenohumeral** joint. In order for the **head of the humerus** to be retained in the glenoid fossa, the muscles doing the retaining must insert around (that is, *in the same plane as*) the head of the humerus and not below it. Muscles inserting on the upper shaft but below the head have the effect of forcing the head out of the fossa when contracting.

All four of the muscles holding the head of the humerus in the glenoid fossa of the scapula arise from the scapula and insert on the **lesser** or **greater tubercles** of the humerus, which are in the same plane as the glenohumeral joint.

Note:
- One muscle inserts on the anterior aspect of the humerus (lesser tubercle): **subscapularis**.
- One muscle inserts at the crown of the greater tubercle: **supraspinatus**.
- Two muscles insert on the posterior aspect of the humerus (greater tubercle): **infraspinatus** and **teres minor**.

A cuff is thus formed—from the lesser tubercle anteriorly over the top (crown) of the greater tubercle and the posterior part of the greater tubercle—a **musculotendinous cuff**. When these four intrinsic muscles of the shoulder contract in unison, the head of the humerus is held securely in the glenoid fossa. The origins of the cuff muscles are made obvious by their names, with the exception of teres minor.

By studying the illustration you can see the cuff muscles inserting on the posterior aspect of the humerus are lateral rotators, and muscle inserting anteriorly is a medial rotator. Hence the group is often called the **rotator cuff**. The **supraspinatus**, as you can see, is an abductor.

Of the muscles of the cuff, usually only the infraspinatus can be visualized on the surface of the back, and only in well-developed individuals.

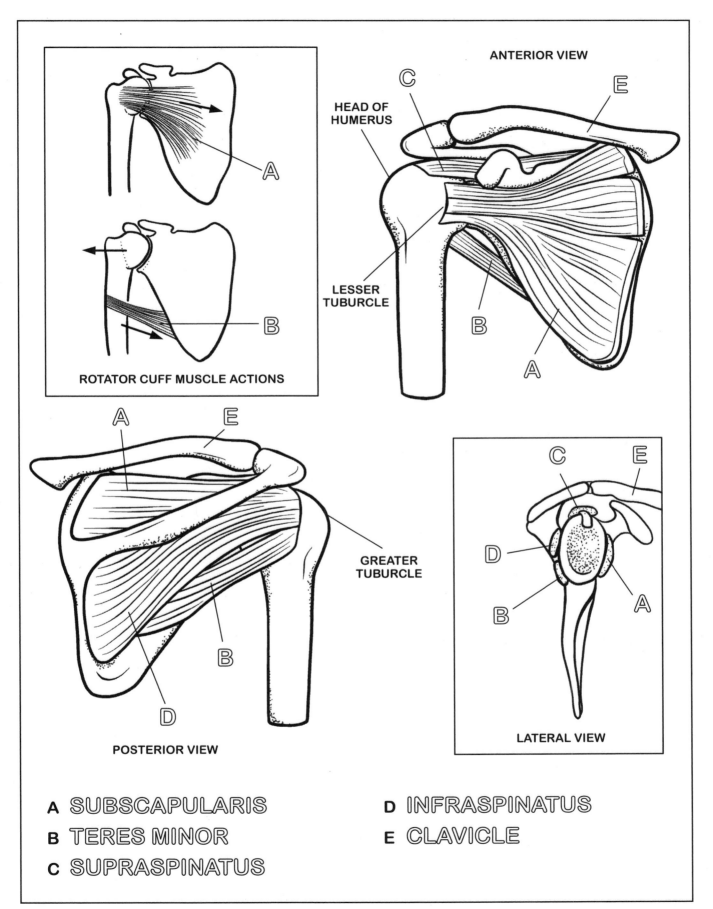

ANTERIOR VIEW

HEAD OF HUMERUS

LESSER TUBURCLE

ROTATOR CUFF MUSCLE ACTIONS

GREATER TUBURCLE

POSTERIOR VIEW

LATERAL VIEW

A SUBSCAPULARIS

B TERES MINOR

C SUPRASPINATUS

D INFRASPINATUS

E CLAVICLE

MYOLOGY OF THE UPPER LIMB
MUSCLES MOVING THE ARM

Pectoralis major is an anterior muscle of the chest, the larger of the two, having three heads of origin: (1) from the **clavicle** (clavicular head), (2) from the **sternum** and ribs (sternocostal head), (3) from the aponeurosis of an abdominal muscle (abdominal head). Attaching to the upper, anterior aspect of the **humerus**, pectoralis major en masse acts to adduct and medially rotate the humerus. You can feel this on yourself. Place your arm (with forearm extended) on the table and turn your body to the side so your arm is abducted. Now adduct your arm against the resistance of the table while placing your hand over your chest. You can probably feel the whole lower border of pectoralis major (in front of the axilla) as well as the upper fibers contract.

The muscle of the shoulder has the shape of the Greek letter *delta*, Δ, but a delta turned upside down: ∇. Thus the name of the muscle is the **deltoid**. The fibers of this exceptionally strong muscle may be divided into anterior, middle, and posterior segments. You can see that the anterior fibers of the deltoid flex the arm, the posterior fibers extend the arm, and the middle fibers abduct the arm. In fact, the deltoid muscle is (1) the primary flexor of the arm, (2) one of two primary extensors of the arm, and (3) the primary abductor of the arm.

The broadest muscle of the back is the **latissimus dorsi**. This muscle is a powerful extensor and adductor of the arm. The lateral border of this muscle can be seen on yourself in a mirror when adducting your arm against resistance from an abducted or extended position. Bring your arm around your chest about a handbreadth below the axilla so that your fingers contract the midback, then extend them so you can feel latissimus dorsi bulge into your hand. This muscle is particularly important in resisting superior displacement of the whole pectoral girdle when a person hangs by the arms or is using crutches for support of the body.

The only other muscle (appropriate to this section) left for consideration is a round muscle, the larger of two: **teres major**. This muscle is an adductor and medial rotator. Teres minor is a lateral rotator.

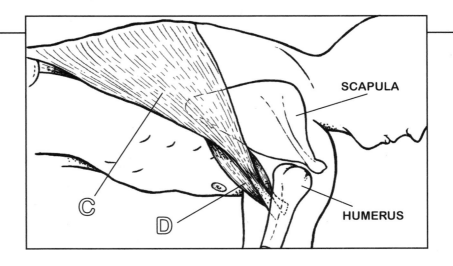

SCAPULA

HUMERUS

C

D

A PECTORALIS MAJOR
B DELTOID
C LATISSIMUS DORSI
D TERES MAJOR

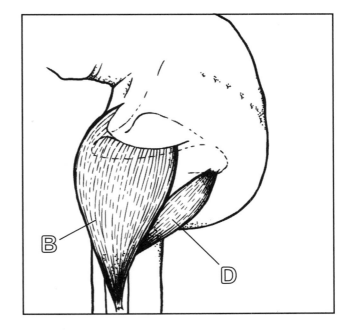

B

D

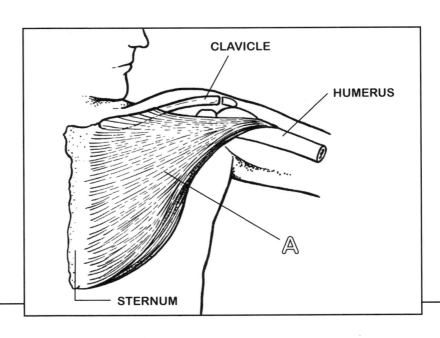

CLAVICLE

HUMERUS

A

STERNUM

MYOLOGY OF THE UPPER LIMB
MUSCLES OF
THE ARM

The muscles of the arm arise from the **humerus** or from the **scapula** immediately adjacent to the humerus, and all but one cross the elbow joint to insert on the ulna or radius. The **humeroulnar joint**, or elbow, is a hinge joint. Hence all the arm muscles on the anterior aspect of the arm are *flexors* of the forearm, and the muscle on the posterior aspect of the arm is an *extensor* of the forearm. The attachments of the principal flexors of the forearm are **biceps brachii** and **brachialis**, and the extensor of the forearm, **triceps brachii**.

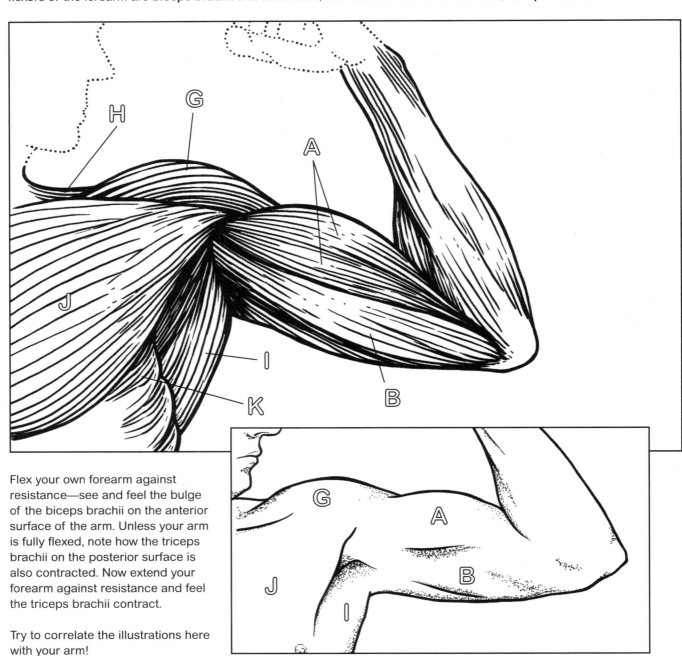

Flex your own forearm against resistance—see and feel the bulge of the biceps brachii on the anterior surface of the arm. Unless your arm is fully flexed, note how the triceps brachii on the posterior surface is also contracted. Now extend your forearm against resistance and feel the triceps brachii contract.

Try to correlate the illustrations here with your arm!

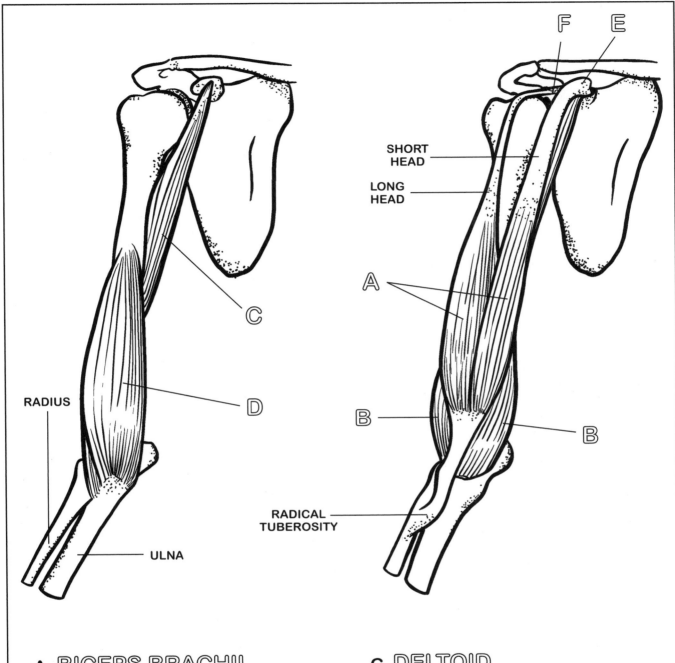

SHORT
HEAD

LONG
HEAD

RADIUS

ULNA

RADICAL
TUBEROSITY

A BICEPS BRACHII

B TRICEPS BRACHII

C CORACOBRACHIALIS

D BRACHIALIS

E CORACOID PROCESS

F SUPRAGLENOID
TUBERCLE

G DELTOID

H TRAPEZIUS

I LATISSIMUS DORSI

J PECTORALIS MAJOR

K SERRATUS ANTERIOR

MYOLOGY OF THE UPPER LIMB
MUSCLES OF
THE FOREARM

There is no shortage of muscles in the forearm—or lack of things for them to do. Consider your own forearm. The bony arrangement is such that two compartments are created: anterior and posterior (with the limb in the anatomical position). Move your hands at the wrist joint and wiggle your fingers. The muscles doing the work, as you can see, are largely in the forearm. Further, the muscles pulling on the anterior aspect seem to be generally involved in flexing the wrist or fingers—thus the anterior compartment might be called the **flexor compartment**. Similarly, the posterior compartment muscles seem to be generally involved in extending the wrist and fingers, thus, the **extensor compartment**.

The principal actors of the wrist found in the forearm are:
• A flexor of the wrist on the ulnar side: **flexor carpi ulnaris**.
• A flexor of the wrist on the radial side: **flexor carpi radialis**.
• An extensor of the wrist on the ulnar side: **extensor carpi ulnaris**.
• Extensors of the wrist on the radial side (a long one and a short one): **extensor carpi radialis longus** and **brevis**. In connection with these names, remember that the wrist is the **carpus**.

Now, flex and extend your fingers without moving your wrist—it is apparent that some muscles responsible for moving fingers are also in the forearm:
• Flexors of the digits, or **flexor digitorum**, are found in a superficial layer (**superficialis**) and deep layer (**profundus**).
• Extensors of the digits (**extensor digitorum**).

The thumb is one of the body's really special structures. It is highly mobile—move it around and see. As you circumduct your thumb (moving it in a circle), note the tendons "pop out" proximal to its base. Note they cross the wrist, and therefore must come off the forearm. Thus these are:
• A flexor of the thumb: longer of two (short one is in the thumb): **flexor pollicis longus**.
• Extensors of the thumb: a long one and short one: **extensor pollicis longus** and **brevis**.
• An abductor of the thumb: longer of two (the short one is in the thumb): **abductor pollicis longus**.

See how you can rotate your forearm (pronation and supination). The principal movers for these functions are also found in the forearm and are reasonably labeled: the **supinator**, **pronators teres** (the round one) and **quadratus** (the rectangular one).

A BRACHIORADIALIS

B PRONATOR TERES

C FLEXOR CARPI ULNARIS

D FLEXOR CARPI MEDIALIS

E PALMARIS LONGUS

F PRONATOR QUADRATUS

G SUPINATOR

H FLEXOR DIGITORUM

I EXTENSOR DIGITI MINIMI

J EXTENSOR DIGITORUM

K EXTENSOR POLLICIS LONGUS

L EXTENSOR INDICIS

M ABDUCTOR POLLICIS LONGUS

N EXTENSOR POLLICIS BREVIS & LONGUS

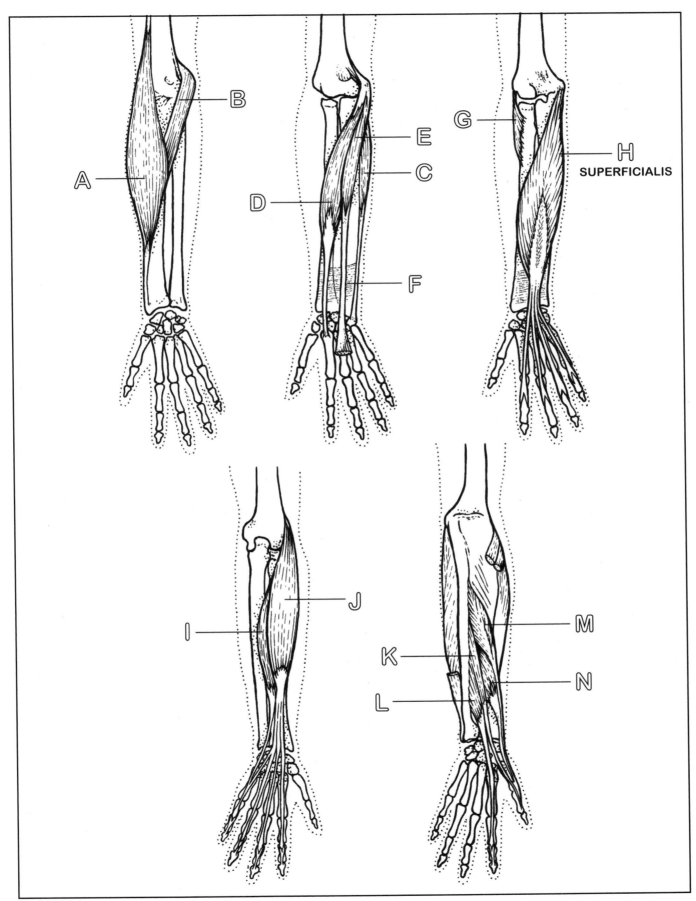

SUPERFICIALIS

MYOLOGY OF THE UPPER LIMB
MUSCLES OF THE HAND

The ultimate anatomical tool is the hand; note the precision with which it conducts its business. Magnificently manipulated by the nervous system, the hand provides the means by which we play piano, create beautiful art, tool a precision valve from a hunk of metal, and perhaps best of all—make new friends with a simple handshake. The scapula, clavicle, humerus, radius, ulna, carpus, and their associated musculature are the servants of the hand; their architectural arrangement you can now appreciate.

Look at the palm of your hand. The muscular pad just below the thumb is the **thenar eminence**. The pad along the little finger side of the palm is the **hypothenar eminence**. In between these pads are small muscles between and anterior to the metacarpals.

The thenar pad is composed of a short abductor of the thumb, **abductor pollicis brevis**, a muscle opposing the thumb to the fifth digit, **opponens pollicis**, and a short flexor of the thumb, **flexor pollicis brevis**.

The hypothenar pad is composed of a flexor of the little finger, **flexor digiti minimi**, an abductor of the little finger, **abductor digiti minimi**, and a muscle opposing the fifth digit to the thumb, **opponens digiti minimi**.

The other intrinsic muscles of the hand include finger adductors (**palmar interossei**), finger abductors (**dorsal interossei**), an adductor of the thumb (**adductor pollicis**), assistant extensors of the interphalangeal joints (**lumbricals, interossei**), and assistant flexors of the metacarpophalangeal joints (**lumbricals, interossei**).

The movement that sets us and primates on a somewhat higher anatomical plateau than our more primitive animal friends with fingers is opposition of the thumb and fifth finger—a rotation of the thumb and flexion of the fifth digit such that the pads of the two fingers come into opposition. This fundamental neuromuscular feat is involved in many of our precise grasping motions.

A PALMER APONEUROSIS

B PALMARIS LONGUS (TENDON)

C PALMARIS BREVIS (MUSCLE)

D ABDUCTOR POLLICIS BREVIS

E OPPONENS POLLICIS

F FLEXOR POLLICIS BREVIS

G FLEXOR DIGITI MINIMI

H ABDUCTOR DIGITI MINIMI

I OPPONENS DIGITI MINIMI

J PALMAR INTEROSSEI

K DORSAL INTEROSSEI

L ABDUCTOR POLLICIS

M LUMBRICALS

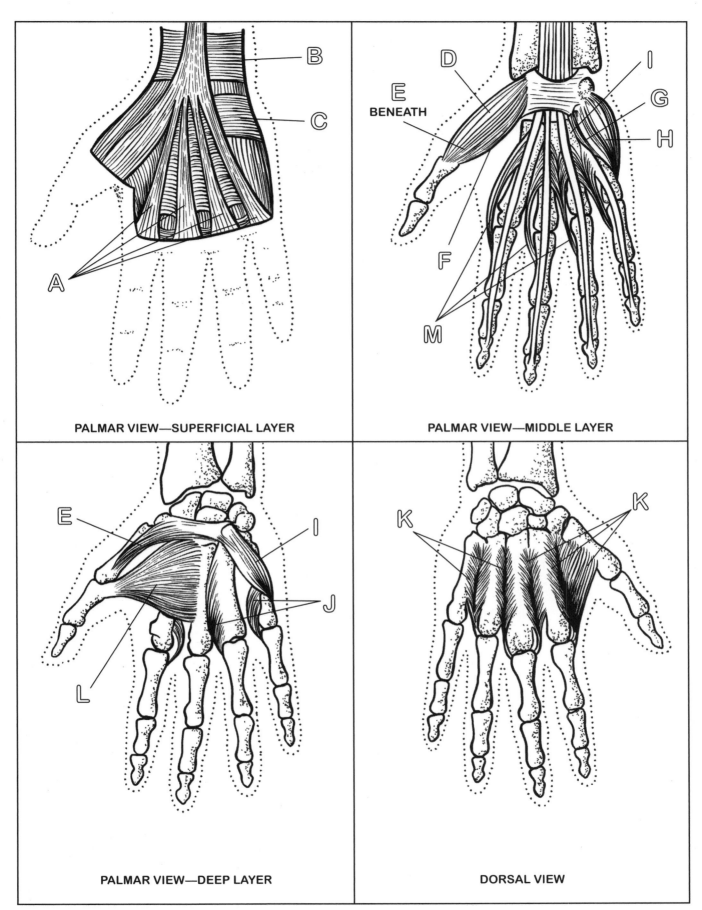

PALMAR VIEW—SUPERFICIAL LAYER

PALMAR VIEW—MIDDLE LAYER

PALMAR VIEW—DEEP LAYER

DORSAL VIEW

REGIONAL & NEUROVASCULAR CONSIDERATIONS: AXILLA

The **axilla** is a space created by the relations of the **scapula**, the **clavicle**, and the **humerus** to the lateral aspect of the rib cage (area of first and second ribs). You will remember that the clavicle forces the scapula laterally and posteriorly. The chest wall is curved laterally and its girth (circumference) decreases from the level of the fifth rib upward. Thus, under the protection of the **acromion**, a space is created.

It is an area of particular interest to the anatomist, clinician, and the surgeon for several reasons, not the least of which is the important neurovascular bundle and its branches/tributaries projecting out to the limb from the neck and chest. Disease of or injury to the axillary structures often involves the scapular region, the breast, pectoral muscles, shoulder, arm, forearm, wrist, and/or hand. Cancer of the breast is often reflected in enlarged lymph nodes in the axilla. In few regions is there such a complex array of structures involving so many areas. Such a fascinating area warrants some investigation.

The axilla is a dome-shaped compartment bounded by:

• An anterior wall created primarily by the **pectoralis major** muscle crossing from the chest to the humerus.
• A posterior wall created by the scapula clothed by the **subscapularis**—plus latissimus dorsi and teres major muscles—which cross the axillary space just lateral to the scapula en route to the humerus.
• A medial wall formed by the second to sixth ribs and intervening intercostal musculature—all of which is covered by the serratus anterior.
• A lateral wall consisting of thin intertubercular grooves of the humerus between the pectoralis major anteriorly and teres major posteriorly.
• The apex represented by the acromion.
• The base created by a suspensory ligament, which gives the axilla a bell shape and is composed of fascia and skin.

Within the space, and bound securely in deep fascia, are:
• The **brachial plexus** of the nerves.
• The **axillary artery** and vein.
• Branches and tributaries of the above.
• The axillary lymph nodes.
• Adipose tissue.

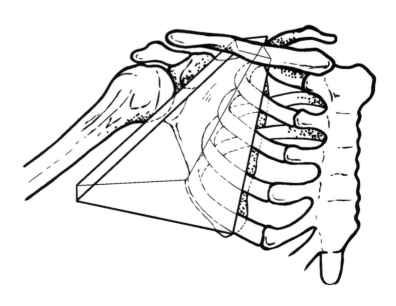

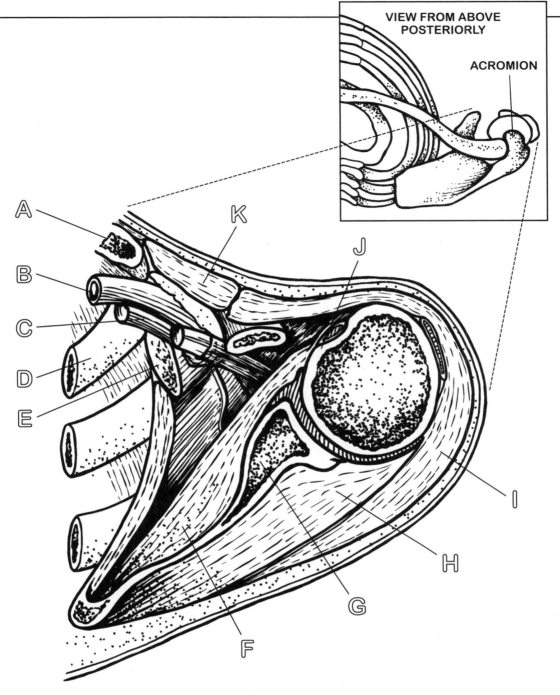

VIEW FROM ABOVE POSTERIORLY

ACROMION

A CLAVICLE

B AXILLARY ARTERY

C BRACHIAL PLEXUS

D FIRST RIB

E AXILLARY SHEATH

F SUBSCAPULARIS

G SCAPULA

H INFRASPINATUS

I DELTOID

J LESSER TUBERCLE

K PECTORALIS MAJOR

REGIONAL & NEUROVASCULAR CONSIDERATIONS
BRACHIAL PLEXUS
& OTHER VESSELS

BRACHIAL PLEXUS

The **brachial plexus** is a network of nerves derived from spinal nerves **C5** to **T1**. It passes out of the neck partly under cover of the trapezius, deep to the clavicle but over the first rib, and into the axilla. Within the axilla the plexus breaks up into an interesting pattern around the axillary artery. From within the axilla, terminal branches develop that supply the entire upper limb with motor and sensory innervation. Because the cords are closely wrapped around the axillary artery they are named according to their relation to that artery: **lateral**, **medial**, and **posterior**. Note on the illustration:

• The roots of the plexus are the anterior rami of spinal nerves C5 to T1.
• These roots soon combine to form three trunks that divide into three cords.

The cords break up into the final branches that pass through the upper limb and supply muscle and skin. Each nerve has sensory and motor components. These terminal nerves are as follows:

• **Axillary nerve** (posterior cord) to the deltoid and teres minor muscles and related cutaneous areas.
• **Radial nerve** (posterior cord) to the triceps brachii, posterior forearm muscles, and related cutaneous areas.
• **Musculocutaneous nerve** (lateral cord) to flexor muscles of the arm and the lateral cutaneous region of the forearm.
• **Ulnar nerve** (medial cord) to the anterior forearm and hand.
• **Median nerve** (from both medial and lateral cords), supplying the lion's share of the anterior forearm and hand.

OTHER VESSELS

The axillary artery is a continuation of the subclavian artery, which itself is a direct branch of the brachiocephalic artery, coming off the arch of the aorta. The branches of the axillary artery supply the shoulder region (including the joints), the pectoral region, and part of the thorax, as well as the proximal arm. The axillary artery continues into the arm to become the brachial artery.

The axillary vein drains into the subclavian vein, which drains into the brachiocephalic vein entering the heart.

There are a number of groups of lymph nodes in the axillary fascia and adipose tissue that are of particular significance: first, because they are the first important group of nodes to receive lymph drainage from the entire limb. Second, because they drain the lymphatic vessels of the pectoral region, i.e., the breast. Thus, any infection of the breast or any part of the arm, forearm, wrist, or hand will pass into the axillary nodes via the various lymphatic vessels and cause their enlargement. Neoplasms of the breast are usually followed by a spreading of the malignant cells (metastasis) into the lymphatic vessels, which transmit the cells to the axillary lymph nodes (as well as to others). These nodes act as a filter in an attempt to prevent further metastasis. Thus early diagnosis of carcinoma of the breast is critical. Palpation of the axillary nodes is commonly performed by physicians during routine physical examinations.

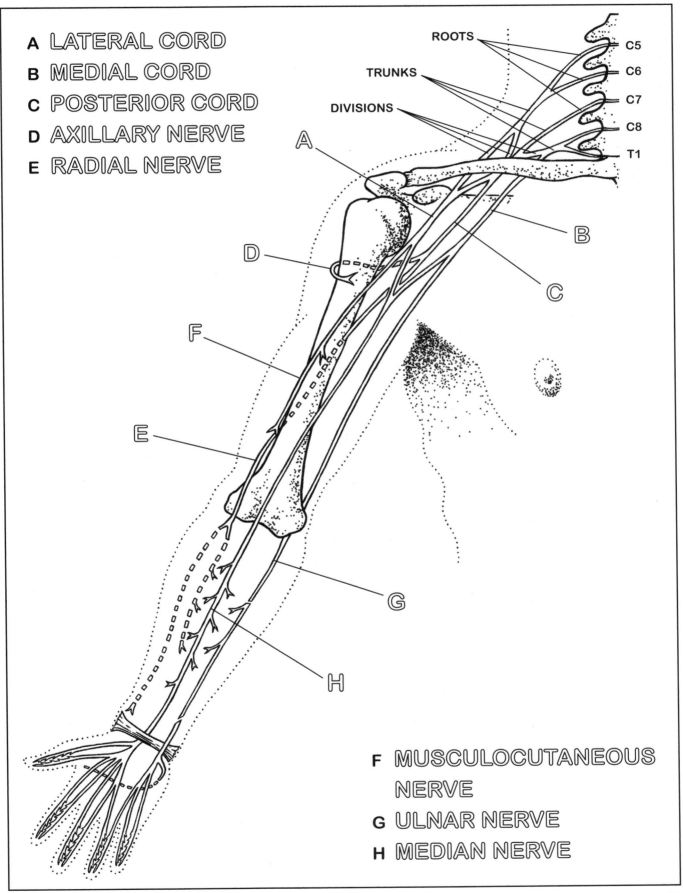

A LATERAL CORD
B MEDIAL CORD
C POSTERIOR CORD
D AXILLARY NERVE
E RADIAL NERVE

ROOTS
TRUNKS
DIVISIONS

C5
C6
C7
C8
T1

A
B
C
D
F
E
G
H

F MUSCULOCUTANEOUS
   NERVE
G ULNAR NERVE
H MEDIAN NERVE

REGIONAL & NEUROVASCULAR CONSIDERATIONS
SCAPULAR REGION

The scapular region, superficial to the rib cage and deep back muscles, is both bony and muscular, consisting largely of the scapula and related muscles: trapezius, rhomboid, levator scapulae, infraspinatus, supraspinatus, and latissimus dorsi. This region boasts a thick but mobile hide. The scapula, trapezius, and latissimus dorsi are normally quite visible and easily palpated in the living person. Most of the muscles (mooring and rotator cuff muscles) are innervated by nerves of the brachial plexus, principally the posterior cord branches. The trapezius is served by the accessory nerve (XI).

The most significant feature of this region is the **scapular anastomosis**—an extensive vascular system and its **anastomoses** (connections between two vessels) arranged in a circular pattern around the scapula, allowing blood to continue to flow in case of damage or pinching of certain scapular arteries. Should the subclavian or axillary artery be **ligated** (cut) following trauma, blood can still reach the most distal part of the limb by these interconnecting vessels.

The source of blood to the scapular region is largely branches of the subclavian artery.

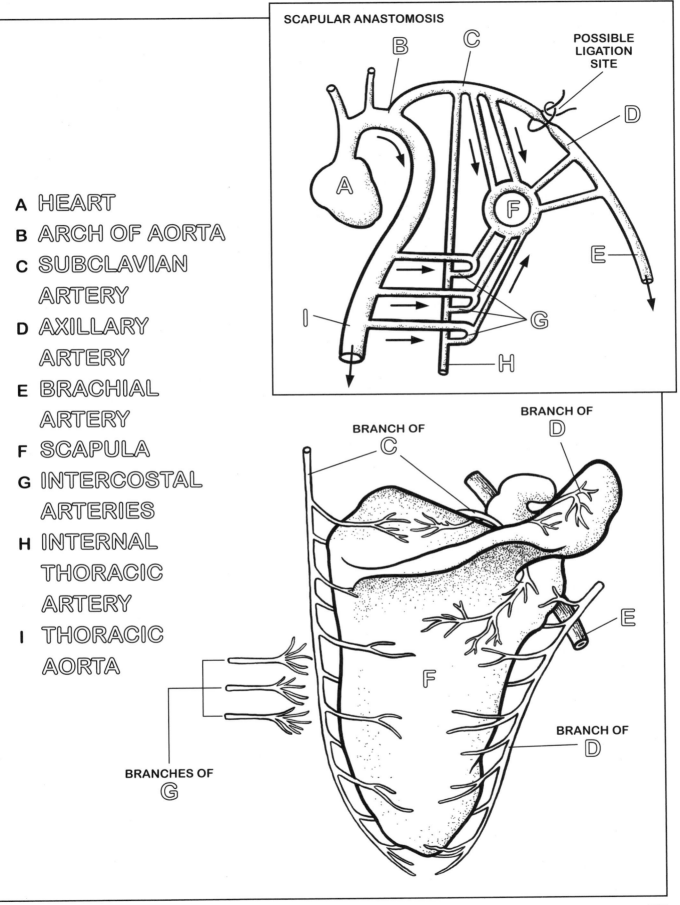

A HEART
B ARCH OF AORTA
C SUBCLAVIAN ARTERY
D AXILLARY ARTERY
E BRACHIAL ARTERY
F SCAPULA
G INTERCOSTAL ARTERIES
H INTERNAL THORACIC ARTERY
I THORACIC AORTA

SCAPULAR ANASTOMOSIS

POSSIBLE LIGATION SITE

BRANCH OF C
BRANCH OF D
BRANCH OF D
BRANCHES OF G

REGIONAL & NEUROVASCULAR CONSIDERATIONS
PECTORAL REGION & BREAST

The anterior counterpart of the scapular region, the **pectoral region**, consists primarily of the **major** and **minor pectoralis** muscles overlying the thoracic cage. Like the scapular muscles, they are concerned with movements of the upper limb. Pectoralis major is easily seen and palpated, and is largely the muscle responsible for the chest form. The pectoral muscles are supplied by the pectoral nerves from the brachial plexus and receive vascular branches from the axillary artery. The skin of this region is significantly thinner than the skin of the upper back, and its underlying superficial fascia is characterized by the presence of mammary glands.

THE FEMALE BREAST

A breast is a group of modified sweat glands, packed in fat, supported by fibrous ligaments borrowed from local fascia, and covered with delicate skin. Underdeveloped in the immature (prepubescent) girl, the breasts begin to enlarge in puberty under the guidance of estrogen, the female sex hormone. The increase in breast size is principally due to a pronounced increase in deposition of adipose tissue. Full development of the glandular and duct components does not take place until pregnancy.

Each breast contains about fifteen to twenty **glandular lobes,** each packaged in generous quantities of **fat.** The **ducts** of the lobes converge toward the apex of the breast, reaching the surface by way of a raised area of skin, the **nipple,** which is supported by connective tissue and smooth muscle. The circular pigmented area of variable size around the nipple (**areola**) is characterized by a number of underlying sebaceous glands. These are active in breastfeeding (lactation) during periods of sucking and function to prevent painful cracking of the skin.

The breasts function to provide nourishing milk for a newborn child. Thus it is not until pregnancy that the breasts fully mature, during which they may double in mass because of the proliferation of glands and ducts.

Blood is transported to the breast by branches of the subclavian, axillary, and intercostal arteries. The pattern of lymphatic vessels of the breast is quite extensive. The lymph from these vessels flows into nodes principally located in the axilla.

The nerve supply is derived from thoracic (intercostal) nerves two to six. These are primarily sensory nerves, for the breasts and nipples are sensitive to touch, which facilitates the expulsion of milk during lactation.

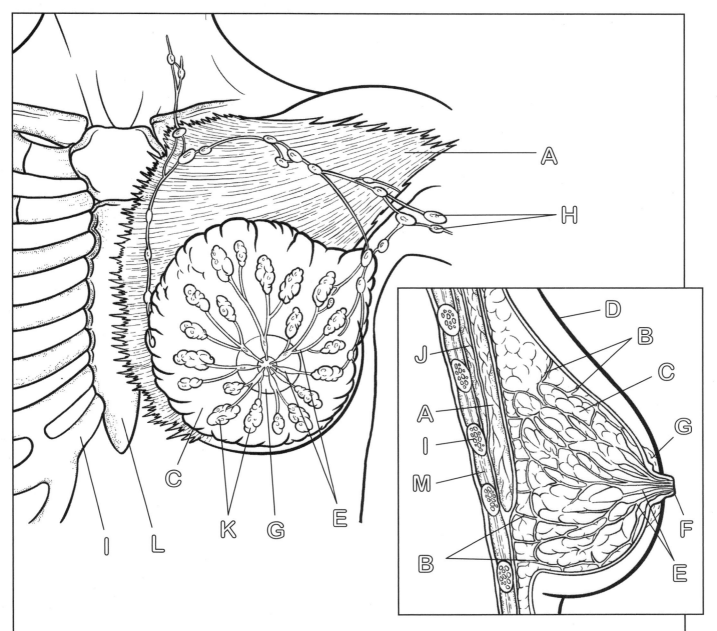

A PECTORALIS MAJOR

B SUSPENSORY LIGAMENTS

C FAT LOBULES

D SKIN

E LACTIFEROUS DUCTS

F NIPPLE

G AREOLA

H LYMPH NODES

I RIB

J PECTORALIS MINOR

K MAMMARY GLAND **(LOBE)**

L STERNUM

M INTERCOSTAL MUSCLES

REGIONAL & NEUROVASCULAR CONSIDERATIONS
SHOULDER & ARM REGIONS

The shoulder region includes the acromion and neighboring scapular spine and lateral clavicle, the acromioclavicular and glenohumeral joints, and related muscles.

The **glenohumeral joint**—a multiaxial, ball-and-socket articulation—is one of the most flexible joints of the body. Its loose ligaments invest a fibrous joint capsule; the socket (glenoid fossa of the scapula) is, as we have mentioned, insecure. A lip of fibrocartilage about the perimeter of the socket helps retain the humerus. Tendons of a number of muscles further reinforce the connection. With numerous muscles approaching and leaving the humerus near the joint at various angles, a great deal of friction is generated in vigorous arm movements—it is no wonder that numerous bursae may be found in and around the shoulder joint where muscles tend to rub against one another (see pages 80–81). These reduce the wear and tear of local muscles and tendons due to friction. Inflammation of these bursae (bursitis) and associated tendons (tendonitis) frequently occurs with strained muscles among laborers and others who constantly put their shoulder joints under stress.

About one and a half inches below the **acromion** on the deep surface of the **deltoid**, the **axillary nerve** and circumflex humeral artery/vein cross from posterior to anterior. Denervation of the deltoid (cutting of the axillary nerve) results in atrophy of the deltoid muscle mass and, in time, a visible shrinking of the shoulder, indicating the extent to which the deltoid gives it form.

The **arm** is the region between the shoulder and elbow joints. In addition to the humerus, it comprises an anterior compartment of principally elbow flexors, a posterior compartment of principally elbow extensors, and a medial neurovascular bundle. All of this can be felt on yourself. Anteriorly, the **biceps** bulges into the hand. Just deep to the bellies of the biceps, lying atop the brachialis is the **musculocutaneous nerve**.

Two other important nerves in the region are the **radial nerve** and the **ulnar nerve**. About one third of the way down the humerus, the radial nerve wraps around the posterolateral humerus in a downward spiraling course en route to the brachioradialis muscle, under which it lies. Thus in arms broken midway up the shaft or higher, the radial nerve may suffer injury. In the lower, lateral aspect of the arm, the brachioradialis muscle can be palpated when the forearm is in the neutral position and the elbow joint is flexed against resistance. The musculocutaneous nerve ducks between biceps and brachialis, and the ulnar nerve heads for the medial epicondyle about one third of the way down the arm. With your fingertips you should be able to roll the ulnar nerve over the medial epicondyle (the "funny bone").

The **elbow** or **humeroulnar joint** is a simple, uniaxial hinge joint between the olecranon of the ulna and the olecranon fossa of the humerus. The medial and lateral epicondyles are easily felt on either side of the joint.

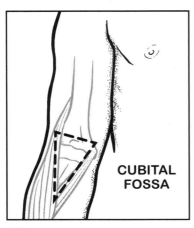

CUBITAL FOSSA

The concavity at the anterior aspect of the elbow is known as the **cubital fossa**, a transition area between the arm and the forearm. It provides a "passageway" for structures to pass between the arm and forearm, such as the **radial nerve**, **median nerve**, and **brachial artery**.

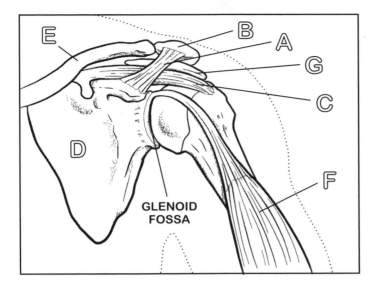

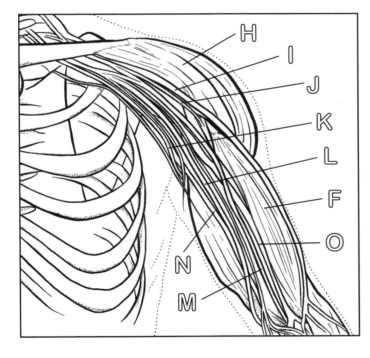

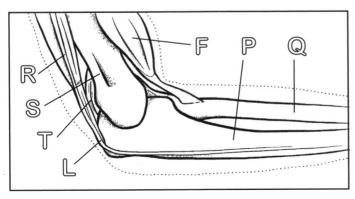

A CORACOHUMERAL LIGAMENT
B ACROMION
C SUBSCAPULARIS MUSCLE
D SCAPULA
E CLAVICLE
F BICEPS BRACHII
G BURSA
H DELTOID
I CEPHALIC VEIN
J MUSCULO-CUTANEOUS NERVE
K RADIAL NERVE
L ULNAR NERVE
M MEDIAN NERVE
N BASILIC VEIN
O BRACHIAL ARTERY
P ULNA
Q RADIUS
R TRICEPS
S HUMERUS
T JOINT CAPSULE

REGIONAL & NEUROVASCULAR CONSIDERATIONS
FOREARM & HAND
REGIONS

As we have seen, the **forearm** consists of the radius and ulna attached to one another at three places (proximal, intermediate, and distal radioulnar joints) and packed tightly with muscle bellies and long sinuous tendons. The bones in intervening interosseous membranes divide the forearm into anterior flex or posterior extensor compartments.

The anterior larger vessels are illustrated here. The **ulnar** and **radial arteries** and **veins**, along with the **median** and **ulnar nerves,** run between the forearm's middle and deep muscle layers. On your own arm, the taut deep fascia make specific muscles difficult to feel, and nerves and vessels are too deep for palpation. But at the interior wrist, the ulnar artery's pulse can be felt on the ulnar side.

The posterior compartment consists of two layers of muscle with the principal neurovascular elements lying between and deep to the two. The superficial layer of muscle is directed to the wrist and digits while the deep layer of muscle is concerned with the thumb and index finger. The **radial nerve** enters the forearm under the brachioradialis, slips under the supinator and terminates in the extensor muscle. Since there are no extensor muscles in the hand (only **extensor tendons**) it follows that the radial nerve will not continue into the hand, except for cutaneous branches.

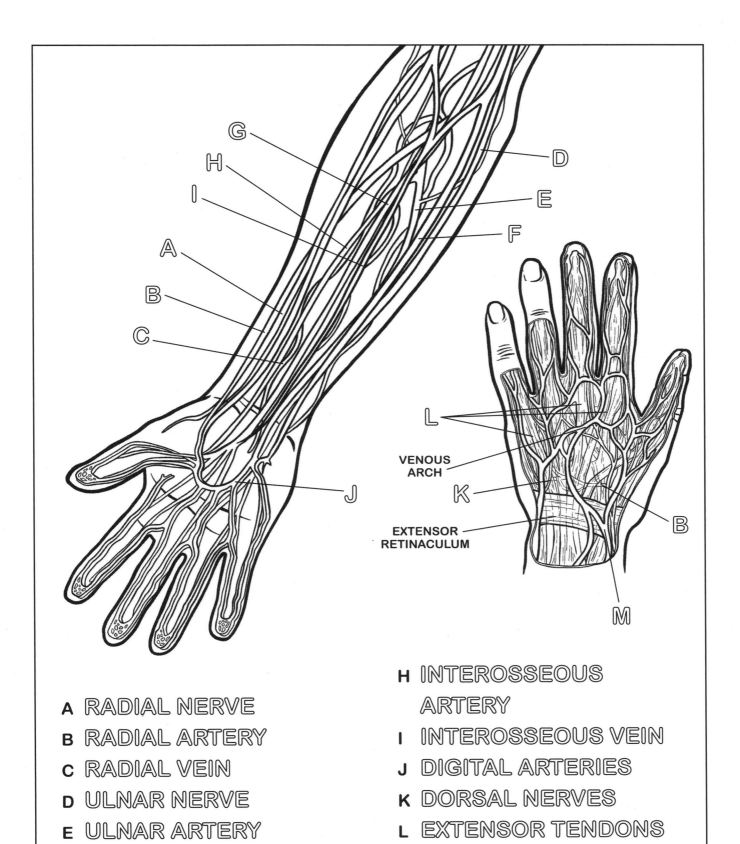

VENOUS
ARCH

EXTENSOR
RETINACULUM

A RADIAL NERVE

B RADIAL ARTERY

C RADIAL VEIN

D ULNAR NERVE

E ULNAR ARTERY

F ULNAR VEIN

G MEDIAN NERVE

H INTEROSSEOUS
ARTERY

I INTEROSSEOUS VEIN

J DIGITAL ARTERIES

K DORSAL NERVES

L EXTENSOR TENDONS

M DORSAL VENOUS
NETWORK

SECTION FIVE:
THE LOWER LIMB

The lower limb is comprised of several complex anatomical parts working together to enable walking, running, jumping, and more. It is home to the largest muscles and the longest bones of the body.

The lower limbs have similar counterparts in the upper limbs. But, unlike the bones of the upper limbs, the lower limbs must support the entire body and provide it with a means of motive power. Their structure reflects these tasks. Specialized for the support of weight, adaptation to gravity, and locomotion, the lower limb consists of several parts or regions: a girdle formed by the hip bones, the thigh, knee, leg, ankle, and foot. In this section, we will take a look at all of them and learn of the entire lower limb.

OSTEOLOGY: BONES OF THE LOWER LIMB

PELVIC GIRDLE

The bones of the lower limb are the rigid framework for the hips, thighs, legs, and feet. They are part of the appendicular skeleton.

The homologue of the pectoral girdle is the **pelvic girdle**, the incomplete circle of bone that supports the vertebral column directly and the body trunk, head, and upper limbs indirectly. The pelvic girdle is composed of two hip bones (right and left) which articulate with one another anteriorly and with the sacrum posteriorly. The pelvic girdle and the sacrum with coccyx make up the circular wall of the pelvis, which it may help to think of as a basin that retains pelvic viscera.

The **pelvis**, via the sacrum, articulates with the fifth lumbar vertebra above and the coccyx below. Anteriolaterally, each hip bone articulates with the thigh bone, the femur.

The **femur** articulates with the larger bone of the leg, the **tibia**, as well as the largest sesamoid bone of the body, the **patella**. This complex three–bone arrangement among the femur, patella, and tibia is the **knee joint**.

The tibia articulates with the femur above, the **talus** below, and the **fibula** at both ends and, by means of a ligament, throughout its length. The tibia and fibula support the soft structures of the leg; the latter bone is long and slender and attaches to the tibia above and talus below.

The **tarsus**, composed of seven bones, supplies the bony support for the **ankle** and posterior half of the **foot**. It articulates with the tibia and fibula (via the talus) proximally and the five **metatarsals** distally.

Each metatarsal articulates with the **phalanx** distally. As in the hand, there are three phalanges in each digit, except for the first digit—which has two.

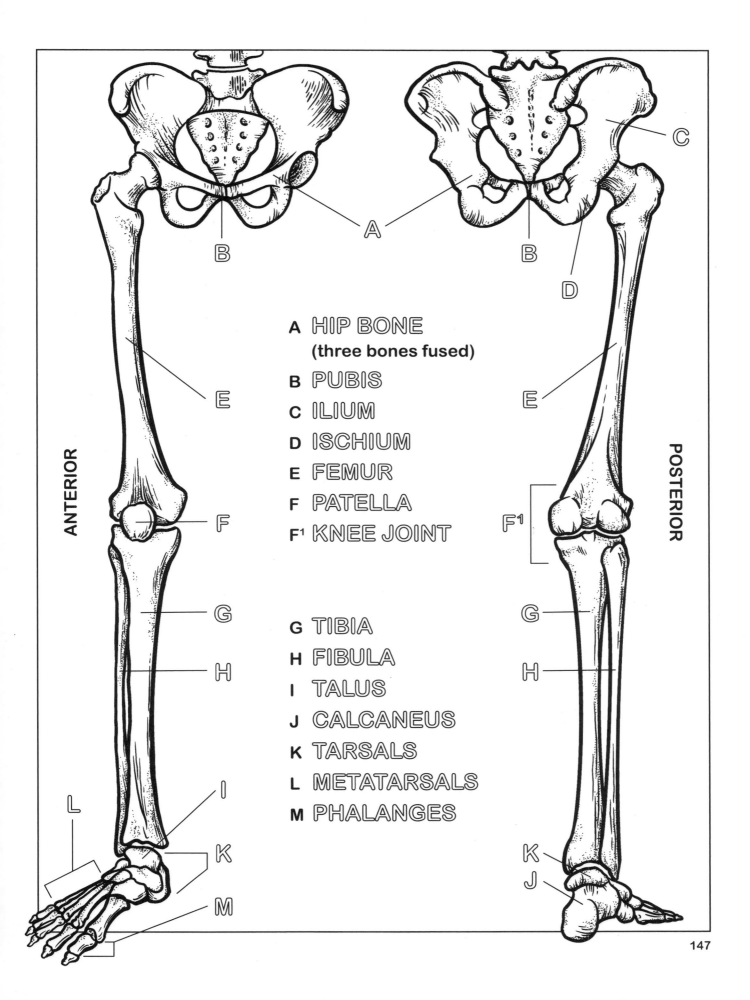

ANTERIOR

POSTERIOR

A **HIP BONE**
(three bones fused)
B PUBIS
C ILIUM
D ISCHIUM
E FEMUR
F PATELLA
F¹ KNEE JOINT

G TIBIA
H FIBULA
I TALUS
J CALCANEUS
K TARSALS
L METATARSALS
M PHALANGES

OSTEOLOGY OF THE LOWER LIMB
HIP BONES

Each **hip bone** articulates with the **sacrum** posteriorly at the sacroiliac joint and with the other hip bone anteriorly at the pubic symphysis.

Until now we've discussed bones that generally lie in a vertical or horizontal plane. A quick look at the skeleton might fool you into thinking that the pelvis is oriented along these two planes the way the main axis of the body is in a standing position. It is important to note that the plane of the upper opening, that is, the **pelvic inlet**, is tilted about 60 degrees from horizontal from posterior to anterior. An appreciation of this orientation is fundamental to an understanding of the following points about the pelvis and its neighbors:

• The abdominal and upper pelvic cavities are without a bony anterior wall.
• The pelvic tilt compensates for the lumbar curvature of the vertebral column.
• The pelvis can handle the weight of the body more efficiently with this orientation.
• There is a tendency for the fifth lumbar vertebra (or its disc) to slip forward on the upper surface of S1. When this happens the condition is called *spondylolisthesis*.

Each hip bone is actually a fusion of three separate bones, the union not taking place until about the age of fourteen. The uppermost bone, characterized by a broad ala (wing), is the **ilium**. It articulates with the other two bones to form the socket, or **acetabulum**, for the head of the femur. The anteriormost of the two is the **pubis**; posteriormost is the **ischium**. Once the centers of ossification within the pubis, ischium, and the ilium join, the bone becomes one, just as a union of the diaphysis and two epiphysis of the humerus make one bone.

Most of the pelvic girdle is covered with muscle, but some features can be palpated. Place your hands on your hips and you will feel a ridge of bone, the **iliac crest**. Trace this crest anteriorly to the point of drop-off. This point is the anterior superior iliac spine, an important attachment point for the anterior thigh muscles.

The pelvic girdle provides support for the pelvic viscera (bladder, rectum, internal organs of reproduction, and associated vessels, nerves, and ducts) and the perineal structures as well (see page 328). The pelvic girdle of females, therefore, must be able to accommodate the developing child during the first nine months of its life and allow it to exit from the uterus within the frame of the pelvic outlet. Since males are without the essential apparatus for such activity, you might suspect that male pelves differ from female pelves. These differences are of course reflected in differences in "typical" male and female body form.

The sacrum and coccyx will be considered with the vertebral column.

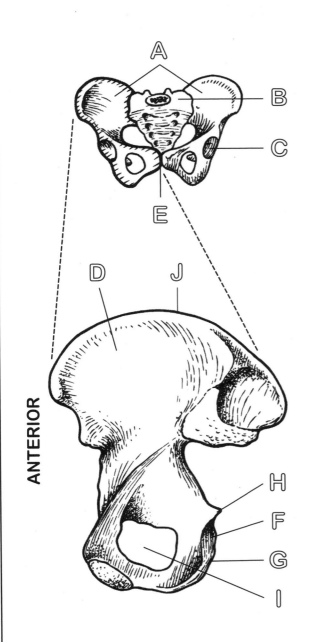

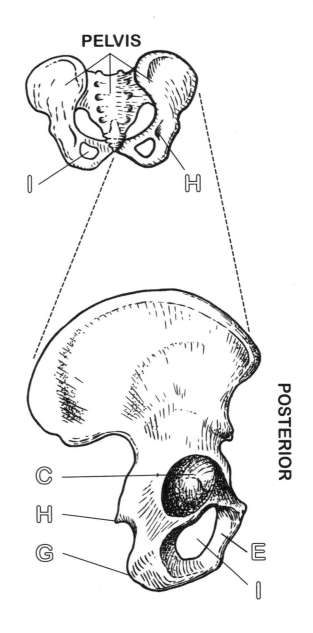

ANTERIOR

POSTERIOR

PELVIS

A | PELVIC GIRDLE

B | SACRUM

C | ACETABULUM

D | ILIUM

E | PUBIS

F | ISCHIUM

G | ISCHIAL TUBEROSITY

H | ISCHIAL SPINE

I | OBTURATOR FORAMEN

J | ILIAC CREST

OSTEOLOGY OF THE LOWER LIMB
FEMUR & PATELLA

FEMUR

The **femur** (Latin for "thigh") is the longest bone in the body. It is the bony framework of the thigh and articulates with the acetabulum above the tibia and **patella** below. Some of the characteristic features of the femur can be felt on your own body. Note the unusual **neck** protruding laterally from the **head**. The neck ends laterally at a large protuberance, the **greater trochanter**, a term implying that there must be a **lesser trochanter**…and there is. Now place the palms of your hands on your iliac crests (lateral aspect) with fingers extending downward. Your distal phalanges should be touching a bony prominence just under the skin and fascia—the greater trochanter. The distance between left and right greater trochanters is greater in females than in males. When one is standing, a horizontal line passing medially from the greater trochanter goes through the center of the hip joint.

The shaft of the **femur** is completely clothed in muscle and cannot be palpated. At the distal extremity of this bone, the **adductor tubercle** can be felt medially just above the medial condyle of the femur. The adductor tubercle is an important attachment point for one of the adductors of the thigh.

The large neck of the femur forces the upper shaft of the bone laterally and so the femur is angled medially from above to below when one stands erect, with knees together. The angle between the neck and the shaft of the femur differs from person to person. In some people in whom the angle is abnormally large, an unusual manner of walking, or gait, is the result.

PATELLA

The **patella** (Latin for "small pan"), or knee cap, is the largest sesamoid bone of the body. Early in fetal development cartilaginous tissue is formed in the tendon of the quadriceps femoris, which crosses the knee joint to insert on the side of the tibia. By puberty (age twelve to fourteen years), the patella has ossified. In the evolutionary sense the patella probably developed in response to trauma of friction generated by the movement of the tendon over the femur (and tibia). The patella articulates only with the femur.

You can easily feel the patella. While sitting extend your knee joint. You should be able to move the patella from side to side and generally trace its entire anterior surface. Note the loose mass just distal to the patella. Keep your hand on it as you slowly flex your knee. That "cord" that takes shape as you stretch it is the patellar ligament, the distal part of the tendon of the quadriceps femoris. Notice that the patella is immobile when the knee is flexed as well as when it is extended against resistance, because the quadriceps femoris—whether stretched or contracted—holds it in place. The patella gives added mechanical advantage to the quadriceps in its job of extending the knee.

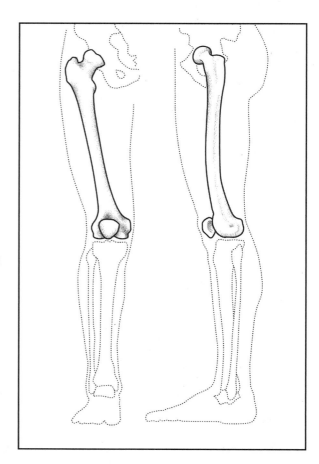

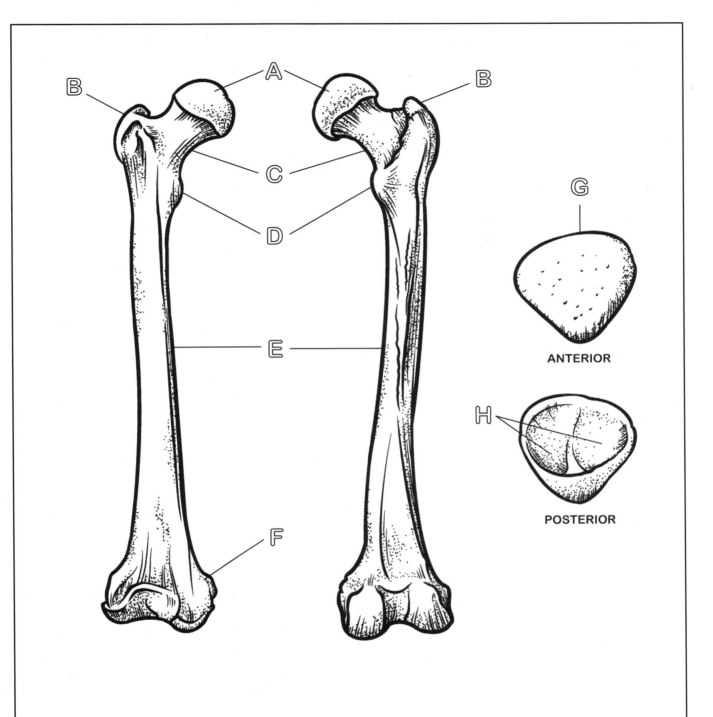

ANTERIOR

POSTERIOR

A HEAD

B GREATER TROCHANTER

C NECK

D LESSER TROCHANTER

E FEMUR

F ADDUCTOR TUBERCLE

G PATELLA

H CONDYLE FACETS

TIBIA & FIBULA

The **tibia** (Latin for "musical pipe"), popularly called the shinbone, is the second-largest bone of the body and corresponds to the ulna of the forearm. The tibia articulates with the medial and lateral condyles of the femur, with the **fibula** throughout on the lateral aspect, and with the talus of the ankle below. The stoutness of the tibia is related to its function—support of the body.

Much of the anterior tibia can be palpated. With the knee flexed, the **medial condyle** can be felt, as can the entire anteromedial surface of the shaft. Note how subcutaneous—that is, how close to the surface—the tibia really is. At the upper end of the shaft you can feel the **tuberosity** of the tibia, which receives the patellar ligament. The anteromedial surface of the tibia terminates distally as the very prominent medial malleolus.

What function might this projection (and its fellow on the lateral side) serve for the tendons from the leg that pass around them and into the foot? Your answer should be: "a pulley."

The fibula is a narrow stick-like bone occupying the lateral side of the leg and corresponding to the radius of the forearm. The fibula not only serves as an important attachment point for muscles but acts as a kind of strut in support of the ankle joint.

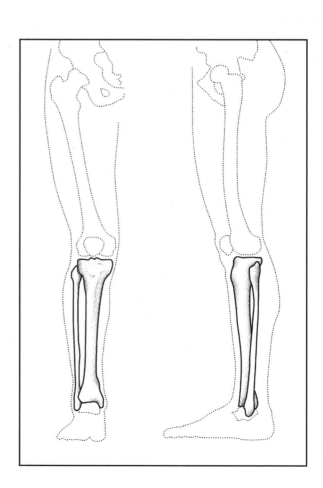

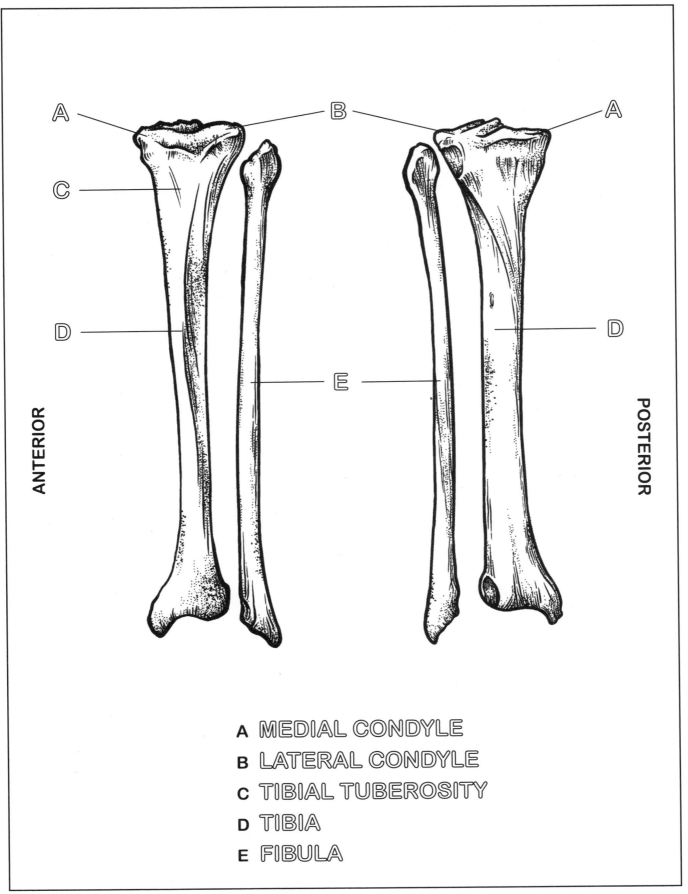

ANTERIOR

POSTERIOR

A MEDIAL CONDYLE
B LATERAL CONDYLE
C TIBIAL TUBEROSITY
D TIBIA
E FIBULA

OSTEOLOGY OF THE LOWER LIMB
BONES & ARCHES OF THE FOOT

If you place your palm down next to the corresponding foot you will see that the bones of the two structures are roughly alike. The bones of the foot consist of the tarsal bones, the metatarsals, and the phalanges.

Approximately one half of the posterior foot is supported by the **tarsal** bones, and you can identify some of these by palpation. The **talus** may be felt about three fingerbreadths below the medial malleolus by approaching the ankle from behind, with your thumb palpating laterally and your fingers palpating medially. Moving forward from below the malleolus about an inch, you will feel a projection on the medial surface: the tuberosity of the **navicular** bone. Feel also the bone of the heel, the **calcaneus**. In the illustration, study the relationship of these three bones. The remaining four tarsal bones, all in line from lateral to medial, are the **cuboid** and three **cuneiforms**. Distally they articulate with the metatarsals, the latter being easily palpated from the dorsal surface.

While sitting down with shoes and socks removed, you should easily be able to feel an arch on the medial side of the foot between the floor and plantar surface. This is one of the components of the **longitudinal arch** of the foot, specifically the medial longitudinal arch. There is also a less obvious lateral longitudinal arch on the lateral side. The longitudinal arch extends from the calcaneus to the heads of the **metatarsals**. The medial component of this arch is formed by:

• Calcaneus
• Talus
• Navicular
• Three cuneiforms
• Metatarsals 1, 2, and 3

Can you confirm this on yourself? If your arch tends to be low, you may not be able to do it. Low arches are quite normal unless they are severe and create difficulty in walking, a condition called *pes planus* ("flat foot"), which is frequently caused by abnormal bone structure or joint defects.

The components of the lateral longitudinal part of the arch include the:
• Calcaneus
• Cuboid
• Lateral two metatarsals

Since the lateral longitudinal arch is the more flat and less mobile one, it is the primary bearer of weight and support, whereas the elastic and flexible medial arch functions in locomotion as a shock absorber and an aid to balance. Both arches are involved in standing and walking. They are maintained by ligaments.

Aside from the longitudinal arch in each foot, there is a transverse arch created when the two feet are placed side by side. The arch flattens when one is standing and adds to the overall flexibility of the foot.

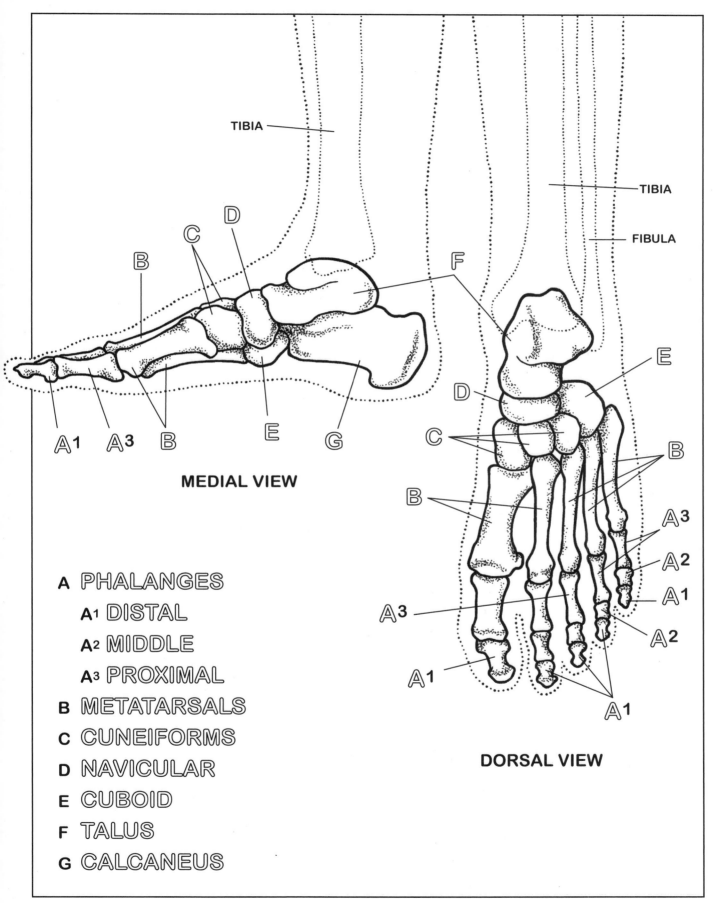

TIBIA

TIBIA

FIBULA

MEDIAL VIEW

DORSAL VIEW

A PHALANGES
  A¹ DISTAL
  A² MIDDLE
  A³ PROXIMAL
B METATARSALS
C CUNEIFORMS
D NAVICULAR
E CUBOID
F TALUS
G CALCANEUS

MYOLOGY OF THE LOWER LIMB: MUSCLES IN THE GLUTEAL REGION

Muscles of the lower limb are numerous but not unmanageable when sorted into groups. Most of them fit neatly into anatomical compartments; further, their names often suggest their function, and their relationships to the joints literally tell the task they perform.

To some degree the lower limb muscles can be correlated with those of the upper limb. However, the correlation breaks down somewhat, partly because lower limb musculature is concerned with support and locomotion, and the upper limb with mobility.

The muscles of the lower limb are organized as follows:
• Muscles of the buttock work the hip joint.
• Muscles of the thigh work the hip and/or knee joints.
• Muscles of the leg work the knee, ankle, and/or foot joints.
• Muscles of the foot work the joints of the digits (toes).

MUSCLES OF THE BUTTOCK (GLUTEAL REGION)

gluteus maximus   **deep lateral rotators**   **gluteus medius**

tensor fasciae   **gluteus minimus**

The muscles of the buttock correspond nicely to those of the shoulder, in a functional sense. Thus lateral and medial rotators, abductors, a flexor, and an extensor can be seen in both regions.

Although there are some ten muscles of the buttock, much of the prominence of the buttock in the muscular individual is created by the massive **gluteus maximus**.

The arrangement of the gluteal muscles is:
• The **deep lateral rotators** occupy an area immediately posterior to the hip joint.
• The **gluteus minimus**, flush against the iliac fossa, lies in the same plane as the lateral rotators, but above them.
• The **gluteus medius** overlies the minimus.
• The thick **gluteus maximus**—most superficial of all—overlies all but a portion of the medius.

The thigh is wrapped in a tight, dense stocking of deep fascia (**fascia lata**). Laterally, the fibers of this fascia are oriented vertically, in a band extending from the ilium to the tibia, which is therefore named the **iliotibial tract**. Arising from the anterior part of the **iliac crest**, a muscle passes down with and inserts into the iliotibial tract. Besides acting to flex and abduct the femur, it also puts tension on the fascia lata and is thus named the **tensor fasciae latae**. Classically, the tensor fasciae latae is included with the gluteal muscles because of its similar innervation. In functional respects, it is more a muscle of the anterior thigh.

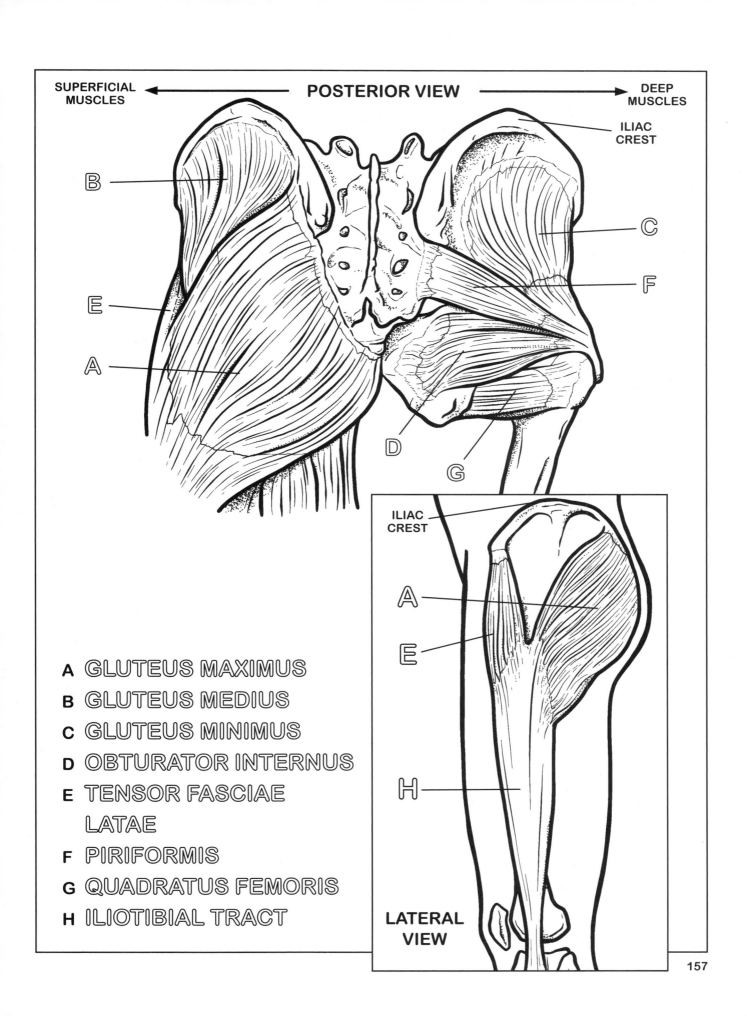

SUPERFICIAL MUSCLES

DEEP MUSCLES

ILIAC CREST

B

C

F

E

A

D

G

ILIAC CREST

A

E

H

LATERAL VIEW

A GLUTEUS MAXIMUS
B GLUTEUS MEDIUS
C GLUTEUS MINIMUS
D OBTURATOR INTERNUS
E TENSOR FASCIAE LATAE
F PIRIFORMIS
G QUADRATUS FEMORIS
H ILIOTIBIAL TRACT

MYOLOGY OF THE LOWER LIMB
MUSCLES OF THE THIGH

The muscles of the thigh are oriented into three compartments.

MUSCLES OF THE POSTERIOR THIGH

semimembranosus   semitendinosus   biceps femoris

These three muscles make up the bulk of the posterior thigh compartment and may be palpated on yourself extensively. The medial tendon is longer and takes up one half of the muscle, which incidentally may help you to remember its name: semitendinosus. Deep to this muscle is the semimembranosus. The tendon palpated on the lateral side is the tendon of a two-headed muscle of the thigh, hence its name: biceps femoris.

Collectively these muscles constitute the "hamstrings," and act together as secondary extensors of the thigh (they cross the hip joint), as well as primary flexors of the leg (they cross the knee joint). These muscles have little slack (are short), as is demonstrated when trying to touch your toes without bending your knees. As the flexion of the hip is increased (e.g., a high kick) the "hams" involuntarily flex the knees.

MUSCLES OF THE MEDIAL THIGH

adductor magnus   adductor longus   adductor brevis   gracilis

These muscles make up the major mass of the medial thigh compartment. As a group they may be palpated when the thigh is adducted against resistance. While standing, spread your legs about two feet apart and hook one foot around the leg of a piece of heavy furniture. Feel now for the inferior pubic ramus and ischial ramus on the side and adduct the limb on that side against resistance. The origin fibers of the adductors may be felt to contract. Feel the extent of this muscle mass inferiorly. This is mainly the adductor magnus. While sitting, spread your thighs and then adduct them against resistance and you can feel as well as see the adductor mass contract. The insertions of these muscles range along almost the entire posterior shaft of the femur. Only the gracilis inserts on the tibia, thus it can be a flexor of the knee as well as an adductor of the femur.

MUSCLES OF THE ANTERIOR THIGH

iliopsoas   vastus medialis   pectineus   vastus intermedius

quadriceps   sartorius   vastus lateralis   femoris   rectus femoris

As we said before, the muscle mass of the anterior side may be functionally subdivided into:
• Muscles that flex the femur into the hip.
• Muscles that flex the femur at the hip and extend the leg.
• Muscles that extend the leg at the knee.

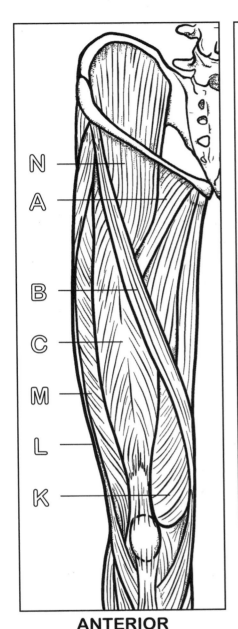

ANTERIOR

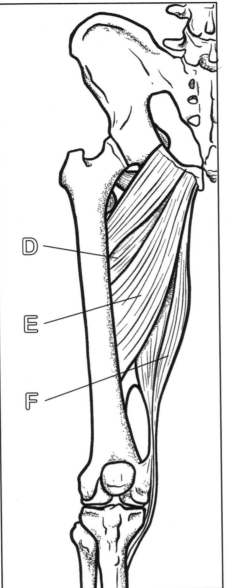

ANTERIOR

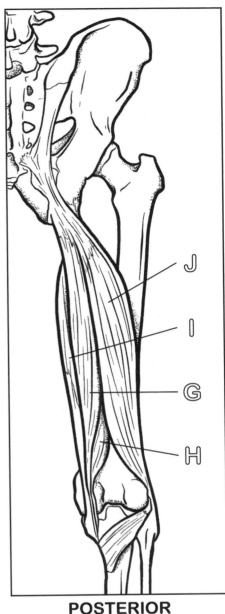

POSTERIOR

A PECTINEUS

B SARTORIUS

C RECTUS FEMORIS

D ADDUCTOR BREVIS

E ADDUCTOR LONGUS

F ADDUCTOR MAGNUS

G SEMIMEMBRANOSUS

H SEMITENDINOSUS

I GRACILIS

J BICEPS FEMORIS

K VASTUS MEDIALIS

L ILIOTIBIAL TRACT

M VASTUS LATERALIS

N ILIOPSOAS

159

MYOLOGY OF THE LOWER LIMB
MUSCLES OF THE ANTERIOR COMPARTMENT OF THE LEG

The muscles of the leg are reasonably named and are therefore easily remembered. You recall the forearm incorporated musculature that manipulated the digits as well as the wrist. So it is with the leg. Further, there are muscles in the forearm that operate the thumb (pollex). Thus there are muscles in the leg that move the great toe (hallux). Like the forearm, the leg contains anterior and posterior muscle compartments. It also has a lateral compartment. The function of these groups is straightforward and is related, in some cases, to the names of the individual muscles.

However, to correlate muscle groups of the leg with those of the forearm, it is necessary to supinate the forearm and hand (as opposed to pronation as in the anatomical position). Place your hands as if they were feet. In this way you can see and remember that:
• Extensors of the foot and toes are in the anterior compartment and are also called **dorsiflexors**.
• Flexors of the foot and toes are in the posterior compartment and are also called **plantarflexors**.
• Muscles (of the anterior and posterior compartments) whose tendons pass to the medial aspect of the foot are inverters/adductors of the foot.
• Muscles (of the lateral compartment) whose tendons pass to the lateral aspect of the foot are inverters/abductors of the foot.

MUSCLES OF THE ANTERIOR COMPARTMENT OF THE LEG

tibialis anterior   **extensor digitorum**   **extensor hallucis longus**

These three muscles of the anterior leg all cross the ankle joint and function to dorsiflex (extend) the foot. The tendon of one of these muscles passes into the medial longitudinal arch and acts on the tarsal joints to effect inversion and adduction.

MUSCLES OF THE LATERAL LEG

peroneal longus   **peroneal brevis**

The peroneal longus and brevis are also known as the fibularis longus and brevis. Peroneal longus is the more superficial of the two and can be seen and palpated on the lateral aspect of the leg when rotating your ankle outward.

MOVEMENT OF THE FOOT
Inversion of the foot is when you tilt the sole of the foot inward at the subtalar joints, when you may stand on the outside edge of your foot. **Eversion** of the foot means to turn the sole of your foot outward.

Adduction rotates the distal end of the foot toward the midline, or toward the center of your body, while **abduction** does the opposite—moving the distal end of the foot away from the center of your body.

Dorsiflexion is the action of raising the distal end of your foot up toward your shin. You employ dorsiflexion when you walk! **Plantarflexion** occurs when the foot points down and away from the leg. You use plantarflexion whenever you stand on tiptoe.

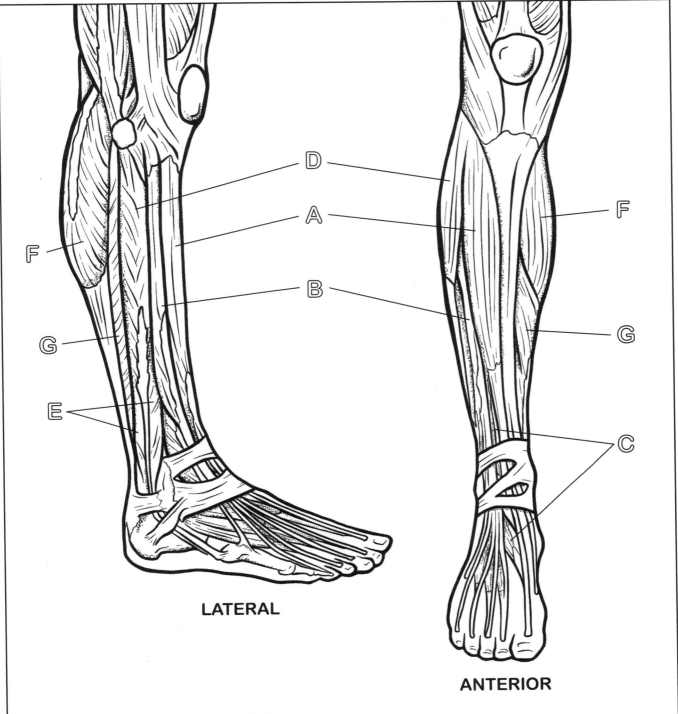

LATERAL

ANTERIOR

A TIBIALIS ANTERIOR

B EXTENSOR DIGITORUM
LONGUS

C EXTENSOR HALLUCIS
LONGUS

D PERONEAL LONGUS

E PERONEAL BREVIS

F GASTROCNEMIUS

G SOLEUS

MYOLOGY OF THE LOWER LIMB
MUSCLES OF THE POSTERIOR COMPARTMENT OF THE LEG

---

gastrocnemius   popliteus   soleus   tibialis posterior

flexor hallucis longus   flexor digitorum longus

Stand on your toes while palpating the back posterior compartment of the leg. If you weigh 200 pounds, that prominent mass you feel just lifted 200 pounds and somewhat precariously set it on the heads of the metatarsals and on your toes. That mass consists of the superficial muscles of the posterior compartment: gastrocnemius and soleus (triceps surae).

Gastrocnemius (Greek, "belly of the leg") and soleus (named after the sole fish, whose shape it resembles) are the principal flexors (plantarflexors) of the ankle joint. They literally pull the heel (**calcaneus**) upward when you stand on your toes, as you have just demonstrated. The muscles narrow distally to form a thick tendon (of **Achilles**) inserting on the calcaneus (tendon calcaneus). "Gastroc," the larger of the two muscles and an important stabilizing muscle of the ankle joint, also crosses the knee joint and is, therefore, a secondary flexor of the leg.

The popliteus, corresponding to the pronator teres in the forearm, acts to flex and medially rotate the leg. It also helps to stabilize the knee joint.

The tibialis posterior is an antagonist of the tibialis anterior in balancing the leg on the foot. Since they both insert on the medial aspect, plantar surface of the foot, they act as inverters/adductors of the foot.

Flexors hallucis and digitorum longus are the antagonists of extensors hallucis and digitorum in movement of the big toe and toes two to five, respectively.

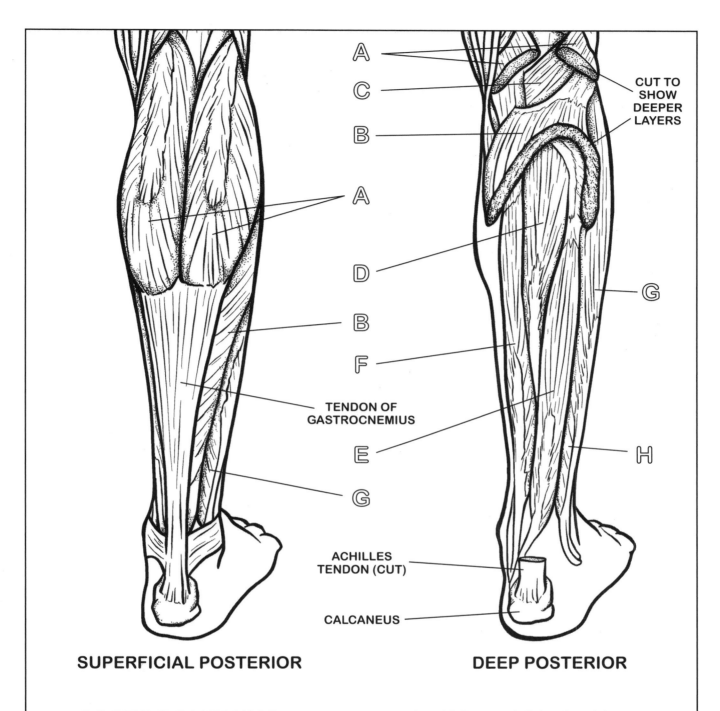

CUT TO SHOW DEEPER LAYERS

TENDON OF GASTROCNEMIUS

ACHILLES TENDON (CUT)

CALCANEUS

SUPERFICIAL POSTERIOR

DEEP POSTERIOR

A GASTROCNEMIUS
B SOLEUS
C POPLITEUS
D TIBIALIS POSTERIOR
E FLEXOR HALLUCIS
LONGUS

F FLEXOR DIGITORUM
LONGUS
G FIBULARIS LONGUS
H FIBULARIS BREVIS

REGIONAL & NEUROVASCULAR CONSIDERATIONS: GLUTEAL REGION

Regionally speaking, the gluteal region is that area of the posterolateral hip bordered by the lumbar region (flank) above and the posterior thigh below. The prominence of the region is the buttock, visible from deposits of fat in the superficial fascia. The left and right buttocks are separated from one another by a gluteal crease; they terminate inferiorly at the gluteal fold.

The muscles of the gluteal region are all movers of the thigh in abduction, rotation, extension, and, in one case, adduction. Secondarily, the deep lateral rotators reinforce the security of the hip joint, as the rotator cuff muscles stabilize the shoulder joint.

BLOOD SUPPLY TO THE GLUTEAL REGION

The blood supply to the gluteal region comes from within the pelvis. The abdominal aorta bifurcates into the common iliac arteries at about the level of the fourth lumbar vertebra. The internal iliac artery (a branch of the common iliac) enters the pelvis and supplies the pelvic viscera. The **superior** and **inferior gluteal arteries** spring directly off the internal iliac, pass through the greater sciatic foramen, and enter the gluteal region to supply the structures there. If the internal iliac artery is ligated, routes of collateral flow may be visualized.

The veins draining the gluteal region ride alongside the arteries, are similarly named, and drain into the inferior vena cava.

NERVES OF THE GLUTEAL REGION

The nerves to the gluteal region arise from the lumbar and sacral segments of the spinal cord via the lumbosacral trunk. The nerves are appropriately named according to the muscles they innervate. Thus gluteus medius/minimus are innervated by the **superior gluteal nerve**; gluteus maximus by the inferior gluteal nerve; **piriformis** by nerve to piriformis.

Passing through the gluteal region is the longest and largest nerve in the body, the **sciatic nerve**, which will innervate the structures of the posterior thigh, the entire leg, and the foot.

A INFERIOR GLUTEAL NERVE & ARTERY
B SUPERIOR GLUTEAL NERVE & ARTERY
C SCIATIC NERVE
D PIRIFORMIS

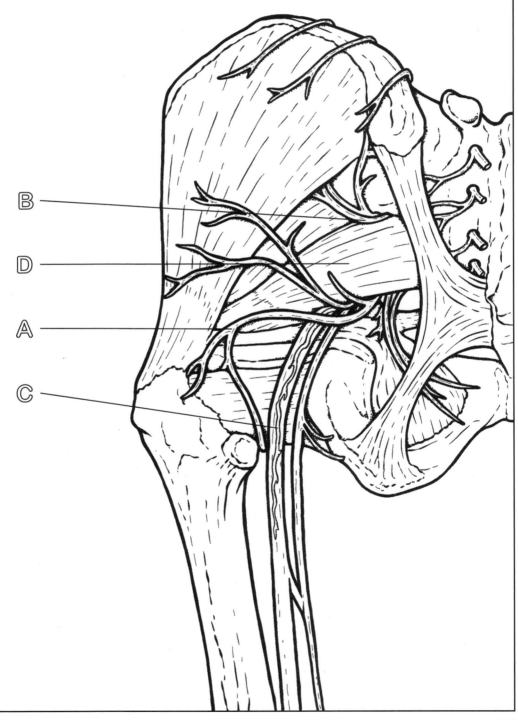

REGIONAL & NEUROVASCULAR CONSIDERATIONS
FEMORAL REGION

The skin and superficial fascia of this region are continuous with those of the buttock. The amount of fat on the thigh is considerably more than that of the arm but less than that of the buttock. The deep fascia of the thigh (fascia lata) is strong and taut. Superiorly it is continuous over the gluteus maximus and is strengthened laterally as the iliotibial tract. It is continuous distally with the capsule of the knee joint.

You may recall that the muscles of the thigh are arranged into anterior, posterior, and medial compartments. In general, each of the three nerves supplying the thigh pass from the pelvic interior directly into its related compartment. The medial thigh compartment (adductors) is served by the **obturator nerve**; the anterior thigh compartment (flexors) is served by the **femoral nerve**; the posterior thigh compartment (extensors of the thigh, flexors of the knee) is supplied by the **sciatic nerve**.

The arterial blood reaches the thigh principally from the **femoral artery,** the distal extension of the external iliac artery. Venous drainage is handled by the **femoral** and **saphenous veins**. The great saphenous vein and its tributaries throughout the thigh, leg, and foot are of clinical interest due to their tendency to become varicosed (abnormally swollen and knotted veins due to sluggish blood flow) in many people.

Lymph flowing through the dense lymphatic network of the thigh and lower limb ultimately drains into the superficial and deep sets of inguinal lymph nodes.

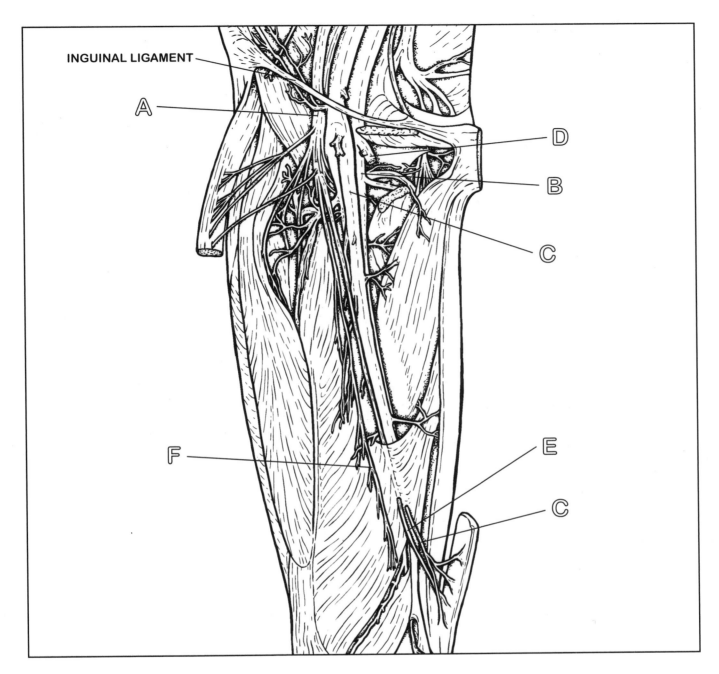

INGUINAL LIGAMENT

A FEMORAL NERVE
B FEMORAL VEIN
C FEMORAL ARTERY
D GREAT SAPHENOUS VEIN (CUT)
E SAPHENOUS NERVE
F MUSCULAR BRANCH OF
   FEMORAL NERVE

REGIONAL & NEUROVASCULAR CONSIDERATIONS
HIP & KNEE JOINTS

HIP JOINT

The hip joint is where the acetabulum of the hip bones articulate with the head of the femur. It is a synovial joint of the ball-and-socket variety. The capsule of the joint is thick and tough, yet loose enough to permit multiaxial movements. The capsule is reinforced by three stout ligaments on all surfaces. Since each ligament arises from one of the bony contributions to the hip, and all attach on the femur, you find the names of these important ligaments quite reasonable. These ligaments are responsible for checking excessive movement at the joint. The interior of the joint capsule is lined with synovial membrane and is partially filled with fat.

The vascular and nerve supply to the joint is by several small twigs from neighboring vessels and nerves.

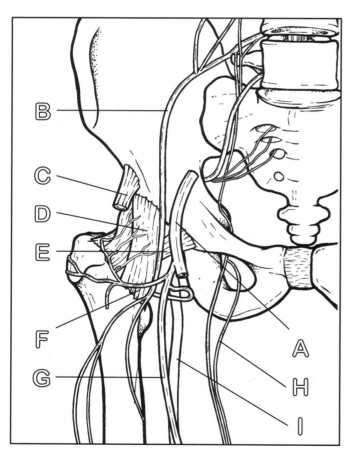

A FEMORAL ARTERY
B FEMORAL NERVE
C RECTUS FEMORIS TENDON
D ILIOFEMORAL LIGAMENT
E PUBOFEMORAL LIGAMENT
F LATERAL CIRCUMFLEX FEMORAL ARTERY
G DEEP ARTERY OF THIGH
H OBTURATOR NERVE
I SCIATIC NERVE

KNEE JOINT

The largest and most complex joint of the body is the combined articulations of the femur with the patella and tibia (at both of its condyles). All three articulations are synovial in character—the patellofemoral articulation is the gliding type and the tibiofemoral articulations are both essentially the hinge type. The knee supports the weight of the body. The center of gravity in the erect position passes through the condyles of the femur. It is strengthened by the presence of many ligaments and tendons as well as muscles crossing the joint.

The capsule of the tibiofemoral joints is reinforced medially and laterally by the tibial and fibular collateral ligaments. The tibial collateral ligament is considered to be the primary stabilizing ligament of the knee. It is subject to tearing under excessive side loads. The anterior and posterior cruciate ("cross-shaped") ligaments—the internal ligaments of the joint—act to steady the tibiofemoral articulations during movement. The tendons of the gracilis, the sartorius, and iliotibial tract reinforce the joint medially and laterally. The anterior capsular wall is interrupted by the patella. Note that the condyles of the tibia and femur do not interlock precisely. Fibrocartilaginous lateral and medial menisci deepen the concavity of the tibial condyles.

The joint is serviced by twigs from the three great nerves of the lower limb and by branches of the arterial anastomoses around the knee.

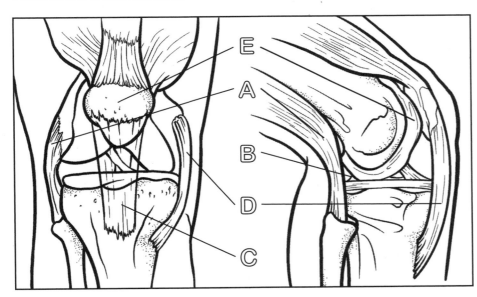

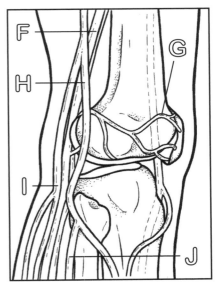

LIGAMENTS:

A LATERAL COLLATERAL
B POSTERIOR CRUCIATE
C ANTERIOR CRUCIATE
D MEDIAL COLLATERAL

E PATELLA
F POPLITEAL ARTERY

G SUPERIOR LATERAL GENICULAR ARTERY

NERVES:

H COMMON PERONEAL
I TIBIAL
J SUPERFICIAL PERONEAL

REGIONAL & NEUROVASCULAR CONSIDERATIONS
POPLITEAL FOSSA

The **popliteal fossa** is created by (1) the popliteal surface of the femur and the posterior wall of the knee joint capsule (both forming the floor of the fossa) and (2) the "hamstrings" above and the gastrocnemius below (these are the fossa's sides). The fossa is covered by muscle, fascia lata, superficial fascia with variable amounts of fat, and skin. The walls of the fossa are easily palpable.

The principal structures in transit within the popliteal fossa are as follows:

- The distal continuation of the femoral artery, the **popliteal artery**. Branches of this artery join with more proximal branches of the femoral and deep femoral arteries to form a collateral route of circulation around the knee.
- The **popliteal vein**, the proximal continuation of the small saphenous, anterior, and posterior tibial veins.
- The **tibial nerve**, one of the two terminal branches of the **sciatic nerve**.
- The common peroneal nerve, the other terminal branch of the sciatic nerve.

All of these structures are, of course, bundled in deep fascia and adipose tissue and may be difficult to palpate because of this. The popliteal fossa is a pressure point in the event of arterial bleeding in the leg or below. A number of popliteal lymph nodes are found within the fossa receiving distal afferent lymphatic vessels and sending off proximal efferent vessels to the inguinal nodes.

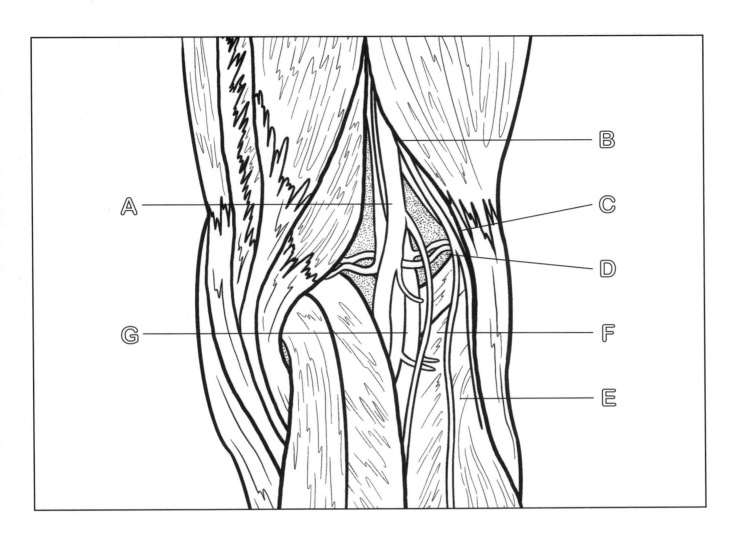

A **POPLITEAL ARTERY**

B **SCIATIC NERVE**

C **COMMON FIBULA NERVE**

D **SUPERIOR GENICULAR VESSELS**

E **SURAL NERVE**

F **TIBIAL NERVE**

G **POPLITEAL VEIN**

REGIONAL & NEUROVASCULAR CONSIDERATIONS
CRURAL REGION

The crural region of the leg (the Latin name for leg is *crus*) lies between the knee and the ankle. As we have seen, it is further divisible into anterior, posterior, and lateral compartments.

The major source of arterial blood to the leg is the **popliteal artery**, which divides into an **anterior tibial artery** and **posterior tibial artery**. The anterior tibial artery courses distally on the anterior aspect of the interosseous membrane, sending muscular branches to the crural extensor muscles. The posterior and lateral compartment of the leg is served by the larger posterior tibial artery whose largest branch, the peroneal artery, helps supply both flexor and extensor musculature.

Innervation to the crural musculature is derived from the sciatic nerve, which splits into the **tibial** and common **peroneal nerves** just above the popliteal fossa. The tibial nerve supplies the posterior compartment; the common peroneal nerve supplies the anterior and lateral regions.

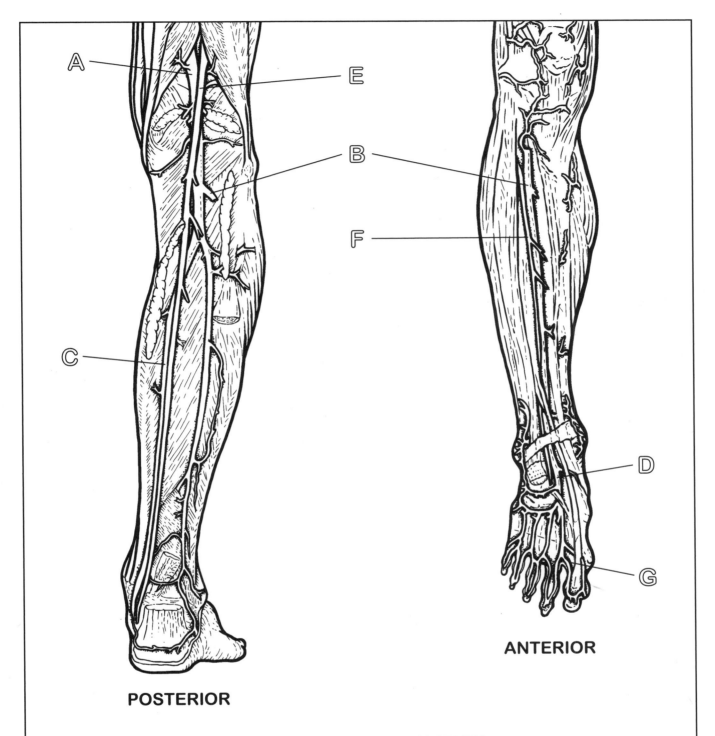

POSTERIOR

ANTERIOR

ARTERIES:

A POPLITEAL ARTERY

B ANTERIOR TIBIAL ARTERY

C POSTERIOR TIBIAL ARTERY

D DORSALIS PEDIS ARTERY

NERVES:

E TIBIAL NERVE

F PERONEAL NERVE

G DEEP PERONEAL NERVE

REGIONAL & NEUROVASCULAR CONSIDERATIONS
FOOT REGION

The region of the foot is generally described as that part of the lower limb distal to the malleoli. The primary support of the foot and the arches of the foot is derived from ligaments. Secondary support is achieved by muscles and the architecture of the bones themselves.

The plantar region of the foot is innervated by branches of the tibial nerve; the extensor region is served by the terminal fibers of the deep peroneal nerve. The arterial supply and venous drainage as well as the nerve supply may be studied on the illustration on the accompanying page.

The tendons of the foot are enclosed in tunnel-like tendon sheaths at critical points, but they lack the interconnections of their fellows in the hand—and as such they are not as subject to injury.

CLINICAL CONSIDERATIONS OF THE LOWER LIMB: NERVE DAMAGE

Injuries to the major nerves of the lower limb can result in reduction in specific movement and use of the affected limb, gait, and weakened extension and flexion of the limb.

Injury to the femoral nerve can be caused locally or may result from trauma to the spinal roots (L2, 3, 4) or the central nervous system itself.

Injury to the obturator nerve, known as **obturator palsy**, is characterized by the affected limb swinging laterally in an arc during walking.

Injury to the sciatic nerve is more common, probably because of its extensive origin and travel through the lower limb. Causes of such injury include diseases of the CNS, a herniated intervertebral disc pressing on the roots of the nerve, stretching of the nerve (as in combined flexion of the hip and full extension of the knee), and other pathological conditions of the pelvis or lower limb.

Injury to the tibial or common peroneal nerve may show up in one's gait.

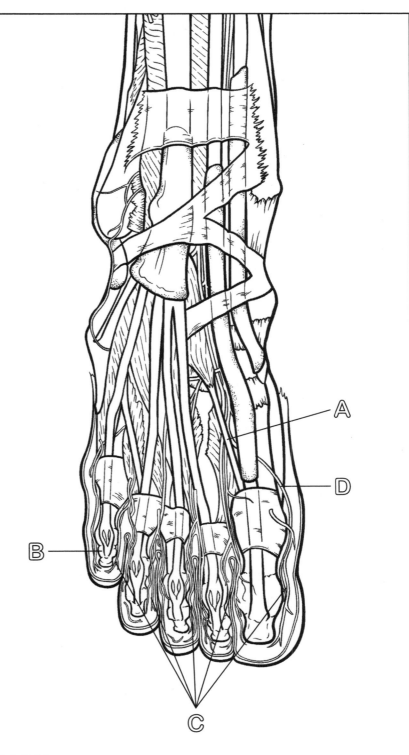

A PLANTAR ARTERY

B SUPERFICIAL BRANCH
OF LATERAL PLANTAR NERVE

C PROPER PLANTAR DIGITAL NERVES

D MEDIAL PLANTAR ARTERY

SECTION SIX:
THE HEAD & NECK

In a manner of speaking, the body is the instrument of the head, and the neck is the thoroughfare between the two. It is in the head that all the decisions are made. It is also there that the air we breathe is first received. It is in the head that we verbalize our thoughts and our feelings, and with some of the organs that we take in food and process it for digestion. Further, the head supports those special sensory receptors related to vision, hearing, equilibrium, taste, and smell, which report directly to the brain. The anterior surface of the head, the face, is so structured with musculature that a myriad of emotional expressions can be created.

It is by way of the neck that orders of movement in response to decisions of the head are carried out, and that the air inhaled reaches the lungs and the food eaten passes to the stomach.

We will now investigate the structure and function of our head and neck.

OSTEOLOGY OF THE SKULL: ANTERIOR ASPECT

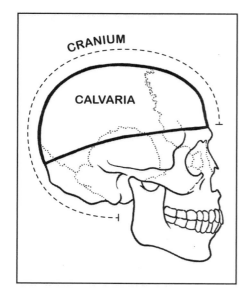

The bones making up the framework of the head are collectively termed the **skull**, with the part forming a vault for the brain being the **cranium**. The vault itself is the **cranial cavity**. The upper, smooth part of the cranium is the **calvaria**, or the skullcap. Its bones are easily palpated—try it! The calvaria includes the **frontal**, **occipital**, right and left **parietal**, right and left **temporal**, **sphenoid**, and **ethmoid** bones. The articulation of the skull with the cervical vertebrae is lost to the touch in the thick and muscular posterior neck.

Most of the bones of the head are interconnected by fibrous tissue, the articulating surfaces being uneven like two pieces of a puzzle. These joints, you may remember, are immovable and are termed **sutures**. With age, the fibrous tissue is replaced by bone (synostosis). In the fetus and newborn infant, the edges of the flat bones of the calvaria have not yet ossified, and diamond-shaped fibrous membranes make up the defect. The "soft spots" of the skull are termed fontanelles, and are eventually replaced by bone.

What remains are the bones of the face, which are also easily felt on the anterior part of the skull. This area is characterized by two **orbits**, the paired nasal cavities, the teeth, and several **foramina** issuing forth some of the arteries, veins, nerves, and lymphatics supplying the facial structures.

The bony nasal cavities cannot be palpated because of the presence of that cartilaginous protuberance (the nose) lengthening the cavities anteriorly. However, the upper segment of the nose (**nasal bones**) can be felt. These two small adjoining bones form the anteriormost bony part of the roof of the nasal cavity.

The oral cavity is a space open to the outside when the lower jaw (**mandible**) is lowered (**depressed**). Note that it is bordered anterolaterally, in closed condition, by the **teeth**. The mandible is easily felt. Start at the "tip" of your jaw and feel upward for the alveolar processes accommodating the lower teeth; note the body of the mandible; feel laterally toward the angle of the mandible.

The bone of the forehead, conveniently termed the **frontal bone**, is the anteriormost bone of the calvaria and protects, in part, the frontal lobe of the brain. Palpate an eyebrow (overlying a supraorbital ridge) and feel for a slight notch on the ridge that may or may not be present—this is the **supraorbital foramen**, accommodating the supraorbital vessels and nerves to the scalp. Within the anterior part of this bone is a cavity, the frontal sinus, which is in communication with the nasal cavity.

The bone taking up most of the facial area and contributing to the orbit and nasal and oral cavities is the right and left halves of the **maxilla**. The maxilla is further characterized by the maxillary sinus cavity (in communication with the nasal cavities), and **infraorbital foramen** below the orbit, through which pass vessels and a nerve of obvious title (infraorbital) to the surrounding face.

The relatively small facial bone separating the maxilla and frontal bone anterolaterally is the **zygomatic**.

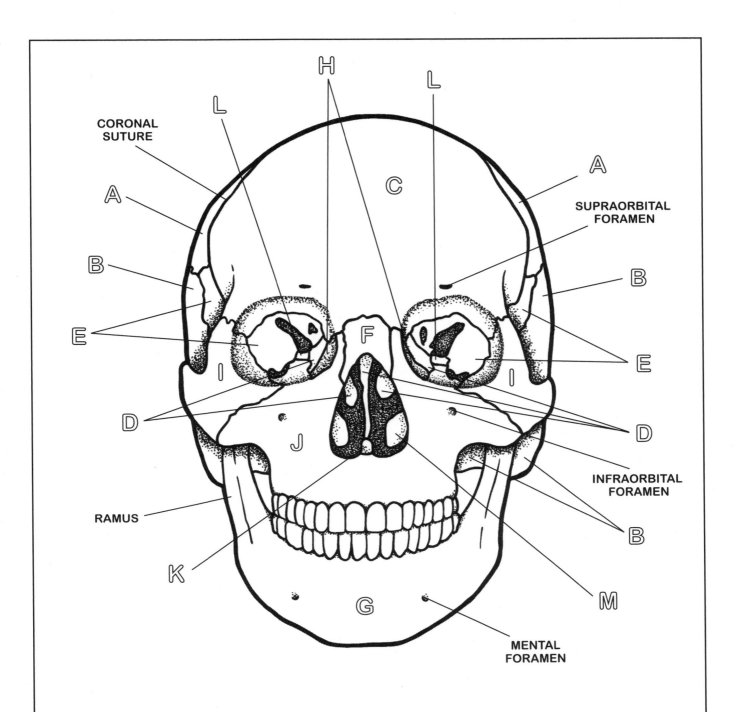

CORONAL
SUTURE

SUPRAORBITAL
FORAMEN

INFRAORBITAL
FORAMEN

RAMUS

MENTAL
FORAMEN

CRANIAL BONES

A PARIETAL
B TEMPORAL
C FRONTAL
D ETHMOID
E SPHENOID

FACIAL BONES

F NASAL
G MANDIBLE
H LACRIMAL
I ZYGOMATIC
J MAXILLA

K VOMER
L PALATINE
M INFERIOR
NASAL
CONCHA

OSTEOLOGY OF THE SKULL
LATERAL ASPECT

The side of the skull is made up of the **frontal** and **parietal** bones, a large part of the **temporal** bone, parts of the **sphenoid** and **zygomatic** bones, as well as the ramus of the **mandible**.

The parietal bones, making up the major portions of the calvaria, protect the parietal lobes of the brain as well as a good part of the frontal lobes.

The temporal bone occupies a major segment of the lateral aspect of the skull. The squamous portion, covered with fascia and muscle, is inferior to the parietal bone and is quite thin. Projecting anteriorly from the squamous portion is the zygomatic process, articulating, appropriately enough, with the zygomatic bone.

Just behind the frontal process of the zygomatic bone is a portion of the sphenoid bone (not palpable, as it is covered with fascia and muscle). It slopes down and inward with this squama of the temporal bone. The projection of the zygomatic and temporal bones (zygomatic arch) skirt around these bones, creating a deep fossa (infratemporal) which is filled with an array of structures—to the anatomist's delight.

The rest of the lateral aspect of the skull is taken up by the ramus of the mandible. Note that the upper border of the ramus (impalpable) ends as two processes. The posterior one is the condyle for the joint; the anterior one is the coronoid process for muscular attachment.

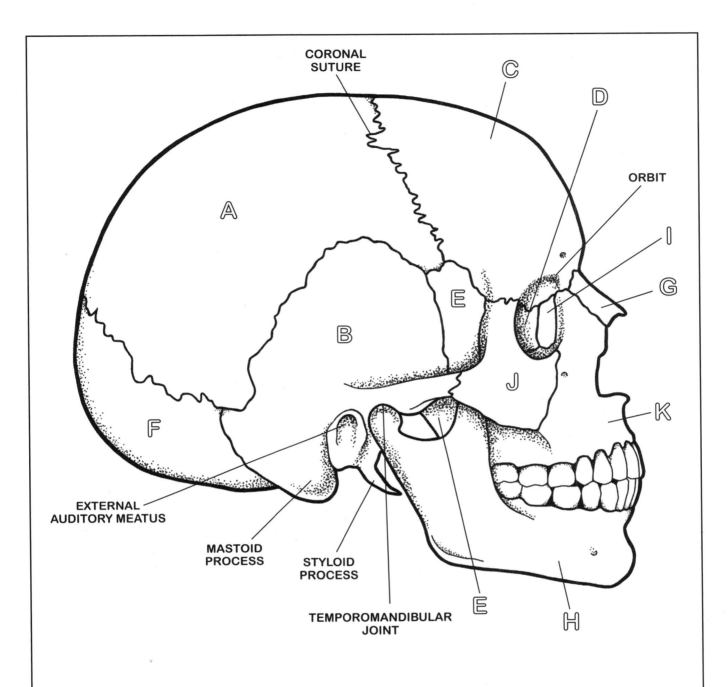

CORONAL
SUTURE

ORBIT

EXTERNAL
AUDITORY MEATUS

MASTOID
PROCESS

STYLOID
PROCESS

TEMPOROMANDIBULAR
JOINT

CRANIAL BONES

A PARIETAL
B TEMPORAL
C FRONTAL
D ETHMOID
E SPHENOID
F OCCIPITAL

FACIAL BONES

G NASAL
H MANDIBLE
I LACRIMAL
J ZYGOMATIC
K MAXILLA

OSTEOLOGY OF THE SKULL
SUPERIOR & POSTERIOR ASPECT

Most of the **posterior** aspect of the skull can be palpated on yourself, including the major part of the **occipital** bone, as well as the contributions of the **temporal** and **parietal** bones. In the middle of the occipital bone, at the upper border of the hollow of the posterior neck, feel for a projection of bone: this is the **external occipital protuberance**. Extending laterally (on each side) from this point are the **superior nuchal lines**, which may not be palpable, as several muscles of the neck and back insert in this area.

The **superior** aspect, or cranial roof, is comprised of the **frontal**, **occipital**, and two **parietal** bones. As we have learned, this is also known as the calvaria. The calvaria is relatively smooth, with small rises and depressions. The anterior border of the calvaria may be felt as the upper ridges of the orbits (bony sockets for the eyeballs); laterally the calvaria disappears deep to the **zygomatic arch**, palpable at the sides of the face; posteriorly it disappears deep to and below the muscles of the posterior neck encroaching upon the **occipital** bone.

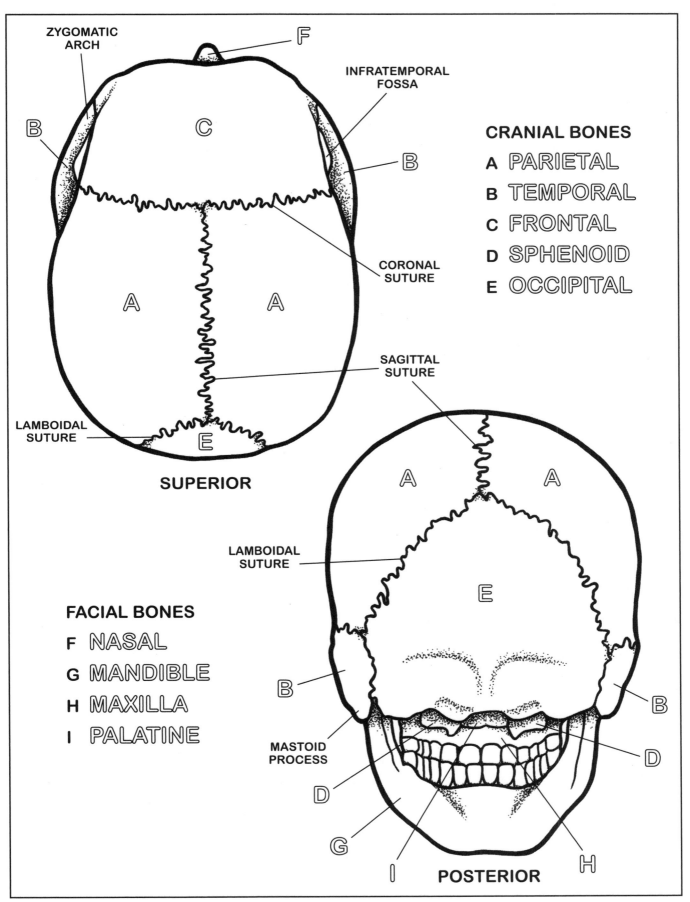

ZYGOMATIC ARCH

F

INFRATEMPORAL FOSSA

B

B

C

CORONAL SUTURE

CRANIAL BONES

A PARIETAL
B TEMPORAL
C FRONTAL
D SPHENOID
E OCCIPITAL

A

A

SAGITTAL SUTURE

LAMBOIDAL SUTURE

E

SUPERIOR

A

A

LAMBOIDAL SUTURE

E

FACIAL BONES

F NASAL
G MANDIBLE
H MAXILLA
I PALATINE

B

B

MASTOID PROCESS

D

D

H

G

I

POSTERIOR

183

OSTEOLOGY OF THE SKULL
INTERIOR

Inside the skull, natural contours provide subdivision into three large spaces called **cranial fossae** (in Latin, *fossa* means trench or ditch). There are three distinct cranial fossae: the **anterior**, **middle**, and **posterior fossa**.

ANTERIOR CRANIAL FOSSA
Supporting primarily the frontal lobes of the brain, the anterior cranial fossa is a high plateau relative to the other fossae. The posterior aspect of the anterior fossa ends abruptly and the middle fossa begins well below the level of the former. The anterior fossa is composed of parts of three bones:

- **Frontal** bone (roof of the orbit)
- **Sphenoid** (lesser wing)
- **Ethmoid** (roof of nasal cavities)

MIDDLE CRANIAL FOSSA
Supporting the temporal lobes of the brain and the pituitary gland (hypophysis), the middle cranial fossa has a central saddle-shaped zone and two lateral cavernous depressions. Many of the nerves and vessels leaving or entering the cranium to and from the neck do so by way of foramina or canals in this fossa—as seen here, the anterior wall of this fossa is the **sphenoid** bone, punctuated by various foramina and fissures, giving passage to nerves and vessels to and from this orbit.

POSTERIOR CRANIAL FOSSA
The posterior cranial fossa—the deepest of the three fossae—supports the cerebellum of the brain. It is formed primarily by the **occipital** bone with an anterior contribution by the petrous (rocky) portion of the **temporal** bone. In this fossa is the large **foramen magnum**, which marks the boundary between the brain and spinal cord.

SUPERIOR VIEW OF SKULL BASE

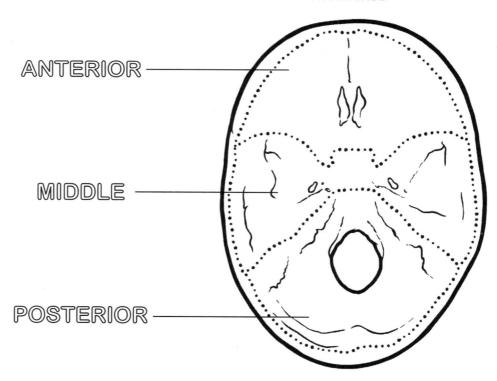

ANTERIOR

MIDDLE

POSTERIOR

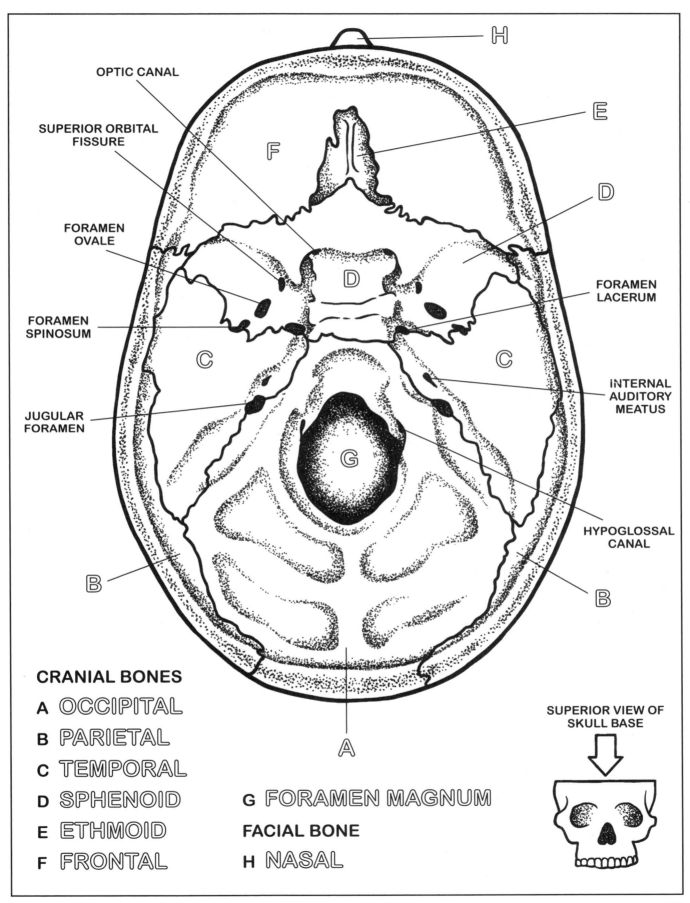

OPTIC CANAL

SUPERIOR ORBITAL FISSURE

FORAMEN OVALE

FORAMEN SPINOSUM

JUGULAR FORAMEN

H

E

D

FORAMEN LACERUM

INTERNAL AUDITORY MEATUS

HYPOGLOSSAL CANAL

F

D

C

C

G

B

B

A

CRANIAL BONES

A OCCIPITAL

B PARIETAL

C TEMPORAL

D SPHENOID

E ETHMOID

F FRONTAL

G FORAMEN MAGNUM

FACIAL BONE

H NASAL

SUPERIOR VIEW OF SKULL BASE

OSTEOLOGY OF THE SKULL
EXTERNAL BASE

The area about the **foramen magnum** represents the **occipital bone**. This bone has a pair of **condyles** lateral to the foramen magnum which articulate with the first cervical vertebra (the atlas). Muscles supporting the head on the neck and providing movement of the head about the neck attach here.

The region between the base of the occipital bone and **palatine** portion of the palate, bordered laterally by the **pterygoid plates** of the **sphenoid**, is taken up by the pharynx and its complex muscular attachments. The pharynx receives air from the nasal cavities, the bony housing of these cavities terminating at the posterior nasal apertures, and the air then passes into the pharynx. The oral cavity also communicates with the pharynx just beyond the posterior edge of the palate.

The region lateral and somewhat posterior to the pharynx is related to a pair of significant **foramina** on each side. The **jugular foramen** transmits the great internal jugular vein and the vagus nerve, along with other structures. The **carotid canal** transmits the internal carotid artery.

The region anterior to the posterior edge of the palatine bone between the upper left and right quadrants of teeth is the bony roof of the oral cavity, the **palate**. The foramina in the area conduct vascular and nerve structures to the mucous membrane of the palate and neighboring areas.

Finally, on the external surface of the base of the skull, there is the **infratemporal fossa**, a space lateral to the attachments of the pharynx and bordered externally by the zygomatic arch. The infratemporal fossa is stuffed with muscles of mastication, muscles relating to the middle ear, and blood vessels and nerves.

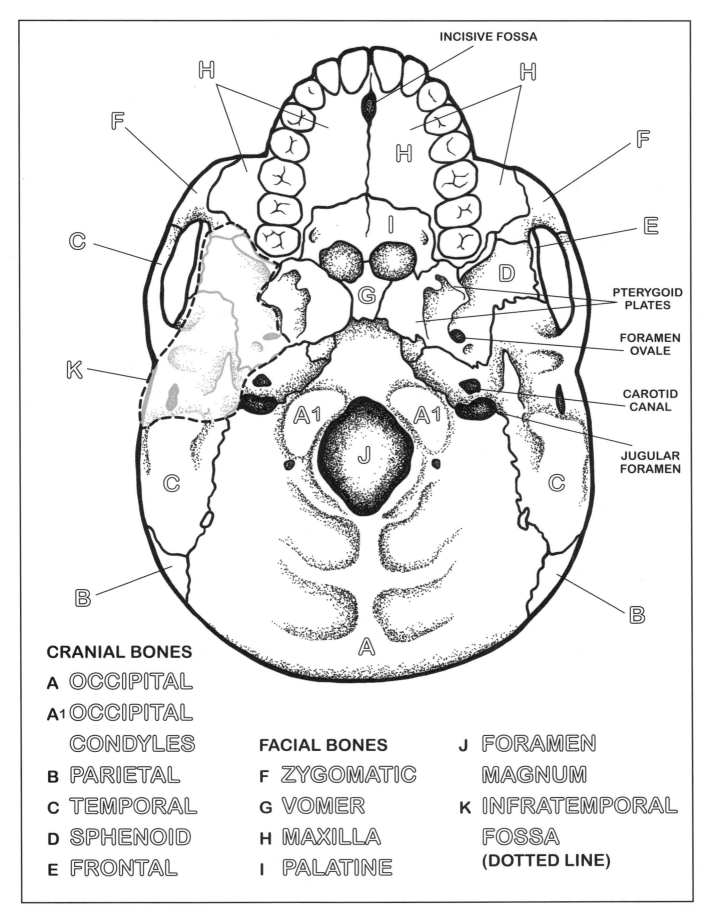

INCISIVE FOSSA

H

F

C

K

B

H

F

E

PTERYGOID
PLATES

FORAMEN
OVALE

CAROTID
CANAL

JUGULAR
FORAMEN

D

G

I

A1

A1

J

C

C

B

A

CRANIAL BONES

A OCCIPITAL

A¹ OCCIPITAL
   CONDYLES

B PARIETAL

C TEMPORAL

D SPHENOID

E FRONTAL

FACIAL BONES

F ZYGOMATIC

G VOMER

H MAXILLA

I PALATINE

J FORAMEN
   MAGNUM

K INFRATEMPORAL
   FOSSA
   (DOTTED LINE)

187

OSTEOLOGY OF THE SKULL
BONES OF SPECIAL SIGNIFICANCE

These are bones whose structure will provide meaningful insight to the organization of the skull. As you study each bone in the following illustration, concentrate on the relationship from one to the other.

SPHENOID BONE

This is the "key" to the skull. It occupies a central position, taking part in the walls of (1) the orbit, (2) the nasal cavity, (3) the outermost two of the three cranial fossae, and (4) the infratemporal fossa. Note that the sphenoid bone is similar to the shape of a bat.

It has a hollow body (sphenoid sinus), articulating with the frontal and ethmoid bones anteriorly and joining posteriorly with the occipital bone. It has a **greater** and **lesser wing** on each side, the lesser wings forming a part of the anterior cranial fossa floor, the greater wings constituting a large part of the posterior wall of the orbit and medial wall of the infratemporal fossa. The space between the two wings is the superior orbital tissue, transmitting vessels and nerves to the orbit. It has a double pair of "legs," the medial and lateral pterygoid plates. A muscle of the pharynx and a pair of masticating muscles attach on these plates.

TEMPORAL BONE

The temporal is a complicated bone that has three basic embryologic parts:

A flat squamous part, which forms a major part of the lateral wall of the skull. The **zygomatic process** projects anteriorly from it. Under the posterior aspect of the process is the **mandibular fossa**, receiving the condyle of the mandible. The temporalis muscle of mastication arises from the squamous part.

A rocky petrous part, separating the middle and posterior cranial fossae. It houses the internal ear (containing vestibular and hearing organs) and middle ear structures. The very important internal carotid artery and facial nerve course through this bone. Laterally, the petrous bone terminates as the **mastoid process**.

A tympanic part, which houses the tympanic membrane (eardrum) and contributes to the **external auditory meatus**.

ETHMOID BONE

The ethmoid is quite delicate because of its thin plates and many air cells. It is a complicated-looking bone, yet can be likened to a simple box. The sides of the box are the **medial orbital plates**; the top of the box is formed by the **cribriform plates**. The box has a middle partition composed of a **perpendicular plate**, above which is the **crista galli** and below is the septum partitioning the nasal cavity into two chambers. Each half of the box interior is again divided into a lateral series of spaces (**ethmoid air cells**) and a medial passageway (nasal cavity).

SPHENOID BONE

A OPTICAL CANAL

B LESSER WING

C GREATER WING

D FORAMEN ROTUNDUM

E FORAMEN OVALE

F DORSUM SELLAE

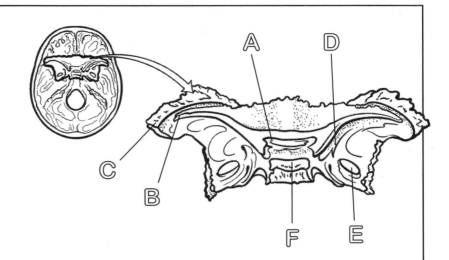

TEMPORAL BONE

G ZYGOMATIC PROCESS

H MANDIBULAR FOSSA

I STYLOID PROCESS

J MASTOID PROCESS

K EXTERNAL AUDITORY MEATUS

L TEMPORAL SURFACE

ETHMOID BONE

M CRIBIFORM PLATE

N CRISTA GALLI

O ORBITAL PLATE

P ETHMOID AIR CELLS

Q PERPENDICULAR PLATE

R SUPERIOR & MIDDLE NASAL CONCHA

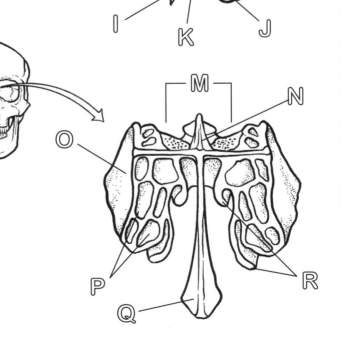

REGIONAL ORGANIZATION OF THE HEAD & NECK: PHONATION & RESPIRATION

There are basically four regions of the head and neck devoted to the generation of intelligible (and unintelligible) noise, and to the preparation of atmospheric air inhaled for oxygenation of the blood by the lungs.

NOSE
The nose consists of the external part of the nasal cavity: a skin-lined cartilaginous affair hung from a roof of nasal and maxillary bone.

NASAL CAVITY
The nasal cavity begins at the nostrils, or **nares**, and extends back to the **choanae**, the apertures that represent the entrance to the **nasopharynx**. It is generally arranged into a vestibule, respiratory region, and upper olfactory region. Behind the nares is the vestibule, where there are stout hairs (vibrissae) and "outdoor" skin gives way to "indoor" mucosa, that is, stratified squamous epithelium is replaced by pseudostratified columnar epithelium with cilia and goblet cells—typical respiratory epithelium. The lateral walls of the cavity are characterized by three mucous-membrane-lined conchae. The underlying meatuses are perforated by passageways from the sinuses. In summary then, the nasal cavity functions to:

• Warm the passing air (blood vessels).
• Humidify the passing air (glands and heat).
• Trap particulate matter (mucous cells and glands).
• Sweep the debris to the oropharynx (cilia).
• Enable the sense of smell (olfactory region).

SINUSES
The sinuses of the skull may be found in the **frontal** bone, the **sphenoid** bone, the maxillary bones, and the ethmoid bone. These are the **paranasal sinuses**.

PHARYNX
The pharynx is divided into three regions:
• The part associated with the nasal cavity, the **nasopharynx**, which is continuous with…
• The **oropharynx**, associated with the oral cavity. This is in turn continuous with…
• The **laryngopharynx**, which is related to the esophagus and larynx.

LARYNX
The larynx is a cartilaginous, ligamentous, and muscular apparatus in the neck which functions to:
• Generate sound and various frequencies and volumes (phonation).
• Guard the lower respiratory tract by means of the cough reflex.

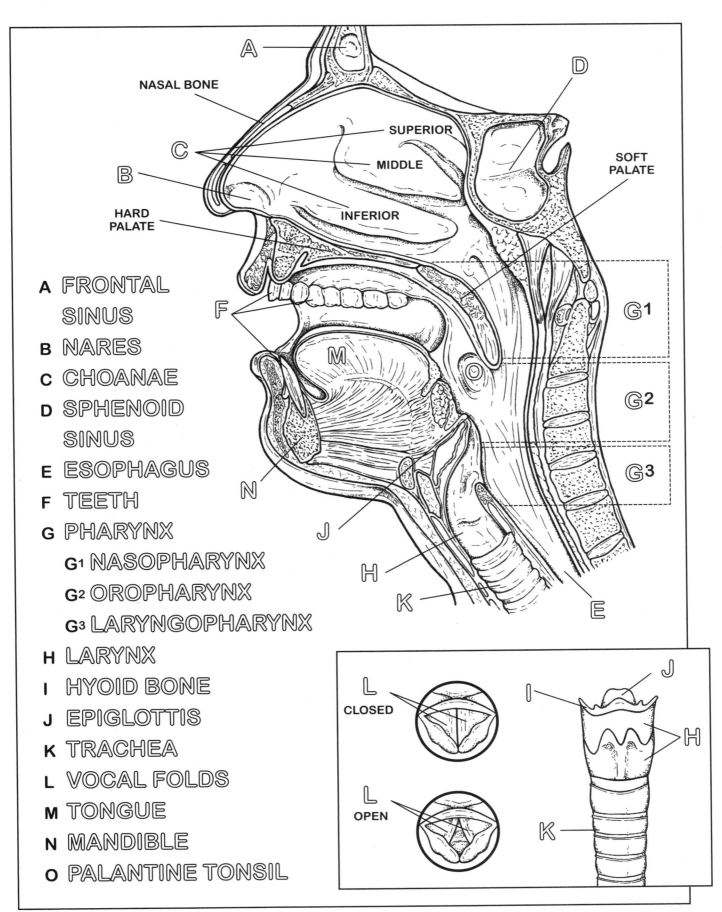

NASAL BONE

SUPERIOR

MIDDLE

D

SOFT PALATE

C

B

INFERIOR

HARD PALATE

G1

F

A FRONTAL SINUS

B NARES

M

O

G2

C CHOANAE

D SPHENOID SINUS

N

G3

E ESOPHAGUS

F TEETH

J

G PHARYNX

  G¹ NASOPHARYNX

  G² OROPHARYNX

H

  G³ LARYNGOPHARYNX

K

E

H LARYNX

I HYOID BONE

J EPIGLOTTIS

K TRACHEA

L

J

L VOCAL FOLDS

CLOSED

I

H

M TONGUE

N MANDIBLE

L

O PALANTINE TONSIL

OPEN

K

REGIONAL ORGANIZATION OF THE HEAD & NECK
THE MOUTH
& MASTICATION

ORAL CAVITY

The structures in the head and neck related to digestion are specifically in the food processing business. Food is initially taken into the body through the **oral cavity** with the aid of the **lips**. Twelve muscles around the lips are the **orbicularis oris**. The lips border the entrance to the mouth, transitioning from skin to a mucous membrane. The mucosa contains no glands, hence the exterior of the lips are dry. Interiorly the lips are glandular and moist. Between the lips and gums (**gingivae**) is the **vestibule**, characterized by a highly glandular mucosa and a midline fold—the **frenulum**. The roof of the mouth cavity is the **hard palate**, lined with glandular, stratified squamous mucosa—and posterior to this, the muscular **soft palate**, terminating posteroinferiorly as the conical **uvula**.

The **floor of the mouth** is largely filled with the **tongue**, whose muscle base is circumscribed by the **mandible**. The lateral walls of the cavity are the cheeks, consisting of the mucosa-lined **buccinator muscle**, covered externally with a fat pad and skin.

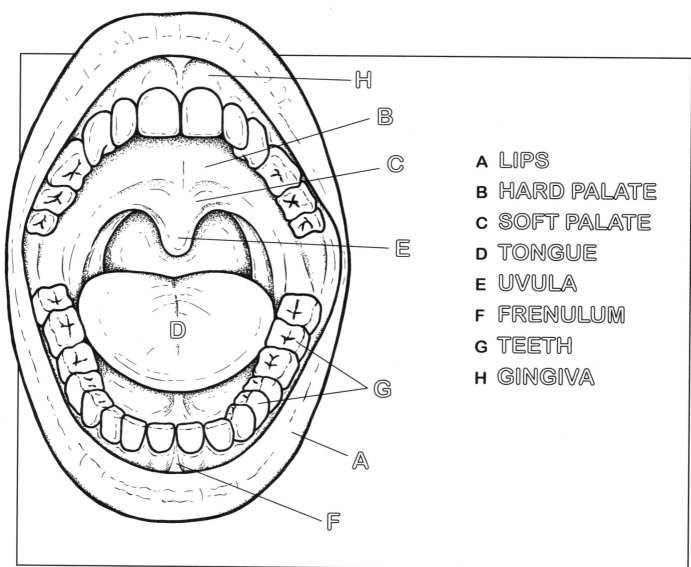

A LIPS
B HARD PALATE
C SOFT PALATE
D TONGUE
E UVULA
F FRENULUM
G TEETH
H GINGIVA

MASTICATION

Food is mechanically processed by chewing, which requires a joint around which the mandible moves, and muscles of **mastication** to move the mandible. The **temporomandibular joint** is the articulation of the head of the **mandible** (condyloid process) with the **mandibular fossa** of the temporal bone. The lower jaw is capable of:

• **Depression** (dropping the jaw down as in opening your mouth).
• **Elevation** (closing the mouth, bringing the teeth into occlusion).
• **Protraction** or jutting of the chin forward. Here the mandible glides forward somewhat in its socket.
• **Retraction** (the reverse of protraction).
• **Lateral displacement** (side-to-side movements).

These movements are generated principally by the four muscles of mastication. The **temporalis** muscle may be felt at the sides of the temporal bone while biting down and the **masseter** muscle can be palpated at the angle of the mandible. The **pterygoid** muscles, residents of the infratemporal fossa, cannot be palpated.

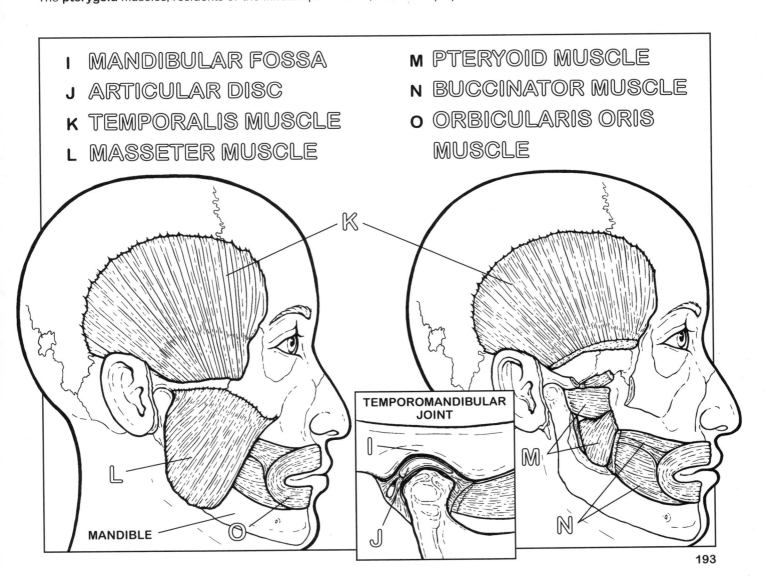

I MANDIBULAR FOSSA
J ARTICULAR DISC
K TEMPORALIS MUSCLE
L MASSETER MUSCLE

M PTERYOID MUSCLE
N BUCCINATOR MUSCLE
O ORBICULARIS ORIS MUSCLE

MANDIBLE

TEMPOROMANDIBULAR JOINT

TEETH

Mastication also requires **teeth** to crush and grind, increasing the surface area of food to allow a more efficient breakdown by enzymes. Teeth also help with speech and assist in the exterior shape of a face.

The teeth are primary agents in processing food that needs to be bitten off, or **incised**, and pulverized. Teeth that fulfill this function are at the front of the dental arches, while those that grind are found more posteriorly in the arches. There are two U-shaped dental arches: upper and lower (**maxillary** and **mandibular**). If an imaginary median line bisected each arch into halves there would be four quadrants of teeth: Upper right and left, and lower right and left.

Commonly known as baby teeth, **deciduous teeth** are the first set of teeth that form during the embryonic stage of human development and **erupt** (become visible in the mouth) during infancy. They are normally lost and replaced by permanent teeth.

Classically there are thirty-two teeth in the adult, or eight in each quadrant. The dental constitution of each quadrant is identical. Four kinds of teeth can be easily distinguished in each quadrant:
Those that incise or nip: **incisors** (two).
Those that tear: **canines** (one).
Those with two cusps that grind: **bicuspids** or **premolars** (two).
Those with four cusps that grind: **molars** (three).

THE ANATOMY OF A TOOTH
Each tooth consists of several parts:

That part projecting above the gum is the **crown**, the surface of which is an **enamel**. Enamel is 96 percent calcified tissue—the hardest in the body. It is not replaced or repaired naturally after formation, and therefore tends to erode with age.

The constricted part of the tooth is the **neck**, where the **gum** attaches. The gum is a mass of thick, dense, fibrous tissue covered with a vascular, pain-sensitive, mucous membrane. Clinically, each gum is called a gingiva; hence inflammation of the gums is **gingivitis**.

The **root** is the part of the tooth embedded in the alveolar socket of the bone, and covered by a layer of fibrous tissue (cement), adhering the tooth to the neighboring bone. The cavitated core of the root and the crown, the **pulp cavity**, contains highly vascular and pain-sensitive loose connective tissue (pulp). Pulp secretes a calcified tissue (**dentin**) which lines the pulp cavity and makes up most of the tooth mass.

Continuous with the pulp cavity, the **root canal** transmits **nerves** and **vessels** to the pulp and adjacent tissues.

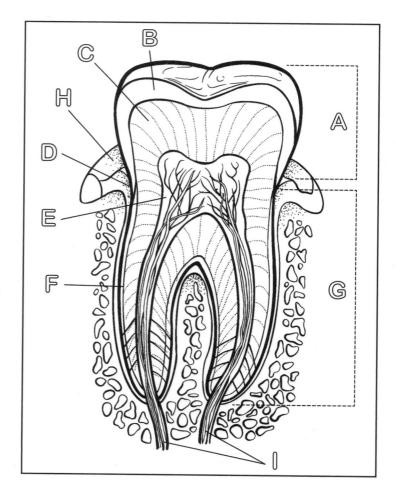

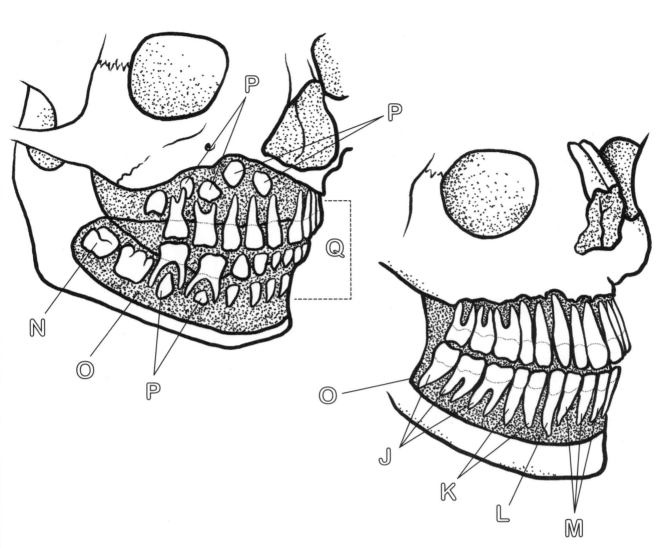

A CROWN		**J** MOLARS	
B ENAMEL		**K** PREMOLARS	
C DENTIN		**L** CANINE	
D NECK		**M** INCISORS	
E PULP CAVITY		**N** THIRD MOLAR	
F CEMENTUM		**O** ALVEOLAR BONE	
G ROOT		**P** DEVELOPING PERMANENT TEETH	
H GUM		**Q** DECIDUOUS SET (BABY TEETH)	
I ALVEOLAR NERVE & VESSELS			

SALIVARY GLANDS & THE PAROTID & SUBMENDIBULAR REGIONS

The **salivary glands** are responsible for the enzymatic processing and wetting of food prior to swallowing and digestion. There are three and they lie adjacent to or near the oral cavity:
• **Parotid**: in the parotid region.
• **Sublingual**: under the tongue.
• **Submandibular**: bent around the mylohyoid muscle.

The **duct** of the parotid muscle enters the oral cavity opposite the upper second molar and can be felt there. The ducts of the submandibular and sublingual glands open at the floor of the mouth along the sublingual papilla and fold. These glands are innervated by both sympathetic and parasympathetic fibers, resulting in a reflex secretion of water and the enzyme **ptyalin** (which works on carbohydrates) in response to the ingestion or thought of ingestion of food.

THE PAROTID REGION
The **parotid region** is an area of the face just anterior and inferior to the ear flap, below the zygomatic arch, and external to the ramus of the mandible. The chief constituents of this region are:

• The **parotid gland** (the largest of the saliva-producing glands) and its duct (the occlusion of which—caused by viral infection—is termed **mumps**).
• The facial nerve (VII) ramifies throughout the substance of the parotid and supplies all the muscles of facial expression.
• The massive muscle of mastication, the **masseter**, easily palpated when the jaws are clenched.

THE SUBMANDIBULAR REGION
A region surrounded by the bodies of the **mandible** bordered below by the hyoid bone and above by the **tongue**, the **submandibular region** houses the **submandibular** and **sublingual glands**, associated **ducts**, vessels, and nerves, and the suprahyoid muscle group.

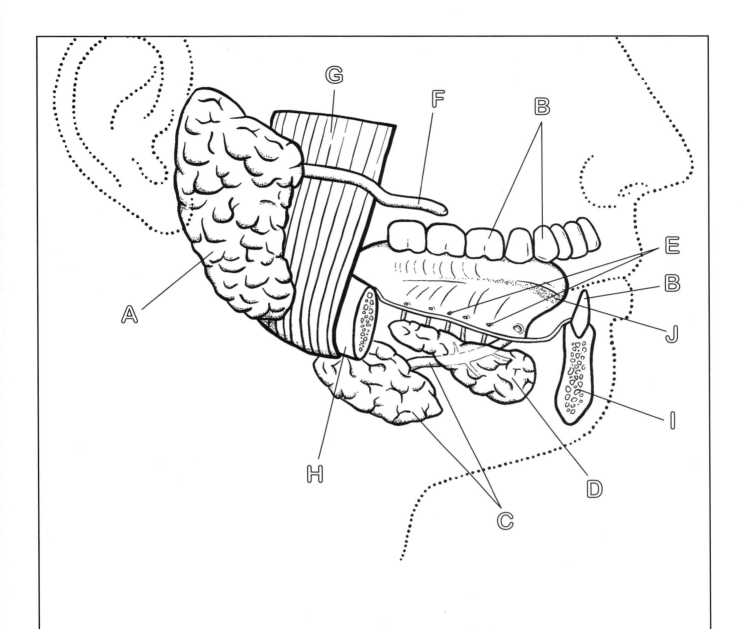

A PAROTID GLAND

B TEETH

C SUBMANDIBULAR
GLAND & DUCT

D SUBLINGUAL GLAND

E SUBLINGUAL DUCTS

F SALIVARY DUCT

G MASSETER MUSCLE

H MANDIBLE **(CUT)**

I ALVEOLAR BONE **(CUT)**

J TONGUE

THE TONGUE & TONSILS

A highly movable collection of skeletal muscles with many nerves covered by a mucous membrane, the main job of the **tongue** is to move food within the mouth while eating. Tongue muscles are covered in a dense layer of connective tissue and finally an outer mucous membrane layer called the **mucosa**. A few different varieties of tiny bumps called **papillae** give the tongue a rough exterior texture. **Taste buds** contain taste receptor cells and are located around the papillae. We will learn more about these taste receptors on page 264.

While many parts of the tongue can perform complex movements, its **root** is firmly anchored to the floor of the mouth. The tongue almost fills the entire oral cavity when the mouth is closed.

The tongue's expert mobility is also used for speaking. When the lips, teeth, and muscular articulation of the tongue work together, sounds from the throat turn into letters and words.

TONSILS

Not just nutrients but bacteria and viruses can also enter the body through the **oral cavity**. Situated on the posterior surface of the tongue and near the entrance of the digestive and respiratory tracts, **tonsils** are tiny organs of lymphatic tissue embedded in the mucosa of the pharynx and are part of the body's immune system. Because of their location they can stop pathogens entering the body through the mouth or the nose with antibody-producing immune cells that destroy germs. Interestingly, a susceptibility to infection does not seem to increase if tonsils are removed. Tonsils begin to shrink as one approaches the teenage years, and will often almost disappear.

Tonsils are oriented around the entrance to the oropharynx in an incomplete ring. The constituents of this ring include the **palatine tonsils**, the **pharyngeal tonsils** or **adenoids**, the **lingual tonsils**, and the **tubal tonsils**.

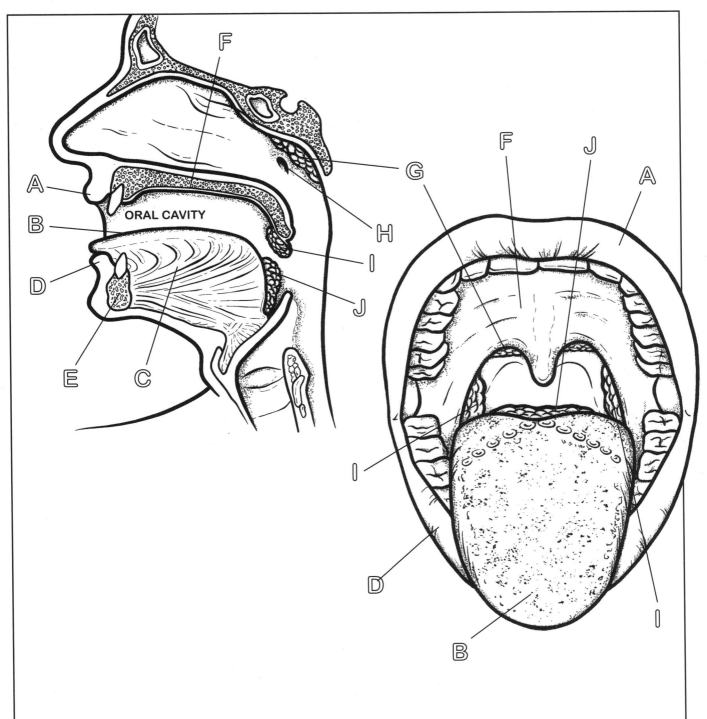

ORAL CAVITY

A UPPER LIP

B DORSUM OF TONGUE

C TONGUE MUSCLE

D LOWER LIP

E MANDIBLE

F HARD PALATE

G PHARYNGEAL TONSILS

H TUBAL TONSILS

I PALATINE

J LINGUAL TONSILS

REGIONAL ORGANIZATION OF THE HEAD & NECK
THE PHARYNX & DEGLUTITION

In the final phase of predigestion, processed food is ejected into the oropharynx by the tongue and swallowed.

PHARYNX

Also called the throat, the **pharynx** is part of the digestive tract that gets food from the mouth to the **esophagus**, which then carries food to the stomach. The muscles constituting the walls of the pharynx are principally constrictors—**superior**, **middle**, and **inferior**. These constrictors are lined internally with mucosa (stratified squamous epithelium) and externally with fascia related posteriorly to the vertebral column.

Laterally, the pharynx is a confluence of many muscles, with contributions from the suprahyoid group, the palate, etc. Inferiorly, the laryngopharynx merges with the muscular tunic of the esophagus.

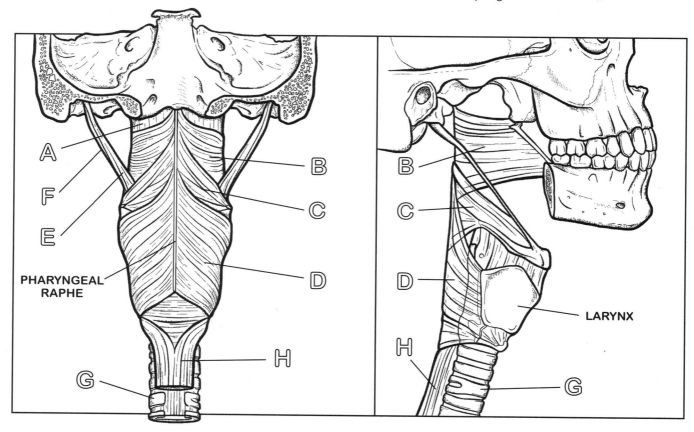

PHARYNGEAL RAPHE

LARYNX

A PHARYNGEAL FASCIA

CONSTRICTOR MUSCLES:

B SUPERIOR

C MIDDLE

D INFERIOR

E STYLOPHARYNGEUS MUSCLE

F STYLOHYOID LIGAMENT

G TRACHEA

H ESOPHAGUS

DEGLUTITION

The act of swallowing or **deglutition** is a rather complex affair, taking place in the pharynx and accomplished by the concerted action of the tongue, muscles of the palate and pharynx, and the esophagus.

Swallowing is divided into three stages:

Stage one: The mouth must be shut. The food mass, or **bolus**, lies on the **dorsum** of the tongue. Inspiration of air ceases. The tongue drives the bolus down through the arches into the oropharynx.

Stage two: The hyoid bone and, therefore, the larynx rise with contraction of the suprahyoid muscles. The **tongue** blocks the oral cavity and the **epiglottis** blocks the **larynx**.

Stage three: The constrictors of the pharynx move the bolus by peristalsis to the esophagus.

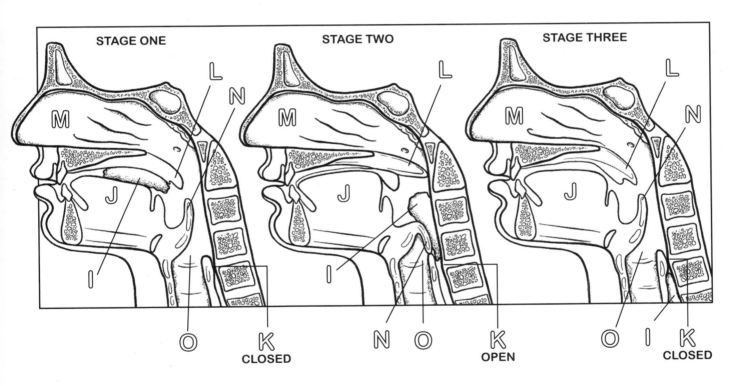

I BOLUS OF FOOD

J TONGUE

K UPPER ESOPHAGEAL SPHINCTER (UES)

L SOFT PALATE

M NASAL CAVITY

N EPIGLOTTIS

O LARYNX

NEUROENDOCRINE SYSTEM: PITUITARY GLAND

The endocrine glands are ductless epithelial structures that secrete hormones into the vascular system. The endocrine glands are generally directly or indirectly influenced by certain nuclei of the hypothalamus that secrete releasing factors. The endocrine system is so tied up with the central nervous system that the functional complex deserves the title **neuroendocrine system**. Components of the neuroendocrine system of the head and neck are the **hypophysis**, the **thyroid**, and the **parathyroids**.

HYPOTHALAMUS

Certainly the **hypothalamus** is a critical component of this system, but its role in the economy of the endocrine system more properly is a topic of physiology, and will not be covered here in any detail. It is sufficient to say that the hypothalamus contains **secretory neurons**, some of whose **axons** terminate in the **posterior lobe** of the hypophysis. The hormones are apparently synthesized in the hypothalamus and released in the posterior lobe, from where they enter the general circulation.

HYPOPHYSIS (PITUITARY GLAND)

The hypophysis, or **pituitary gland**, sits in the saddle (sella turcica) of the sphenoid bone and is connected to the hypothalamus by a **stalk**. The hypophysis may be seen to have an **anterior lobe**, an intermediate part, and a **posterior lobe**. The anterior lobe is known to secrete growth hormone, adrenocorticotropic hormone, thyroid-stimulating hormone, gonadotropins, and prolactin.

Growth hormone (GH) is a protein that stimulates the lengthening of the bone in young people by accelerating cartilage deposition at the epiphyseal plate.

Adrenocorticotropic hormone (ACTH) is a protein that is directed toward the rind (cortex) of the adrenal gland atop the kidney. It stimulates the cortex to secrete a number of hormones called glucocorticoids.

Thyroid-stimulating hormone (TSH) is a protein that influences the thyroid gland in the neck to secrete thyroxine, a hormone that increases the basal metabolism in most normal tissues.

Gonadotropins are a pair of hormones that stimulate the ovary to produce follicles (follicle-stimulating hormone or FSH), and to ovulate (expel the ovum or egg) and transform postovulatory follicles into fatty hormone-secreting bodies (luteinizing hormone or LH). In the male, FSH stimulates spermatogenesis and LH stimulates the secretion of testosterone (male sex hormone) by the testis.

Prolactin is a hormone that stimulates the production of milk in pregnant women.

The posterior lobe is known to release **oxytocin**, an antidiuretic hormone. Oxytocin is a hormone that acts on myoepithelial cells of the mammary glands in response to sucking of the lactating breast causing an injection of milk from the glands and ducts into the mouth of an infant. Oxytocin also stimulates uterine muscle contraction at the time of birth. Antidiuretic hormone or ADH acts primarily on the kidney tubules to retain water.

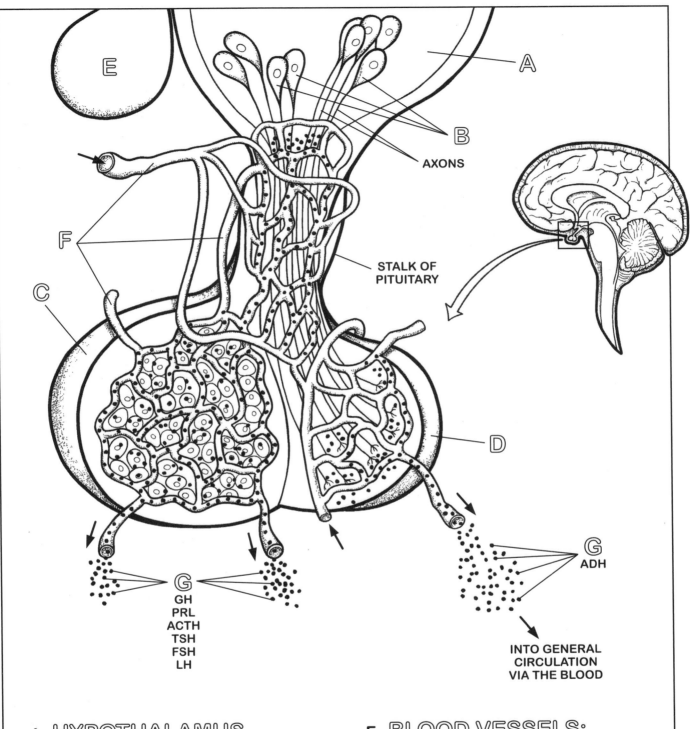

A **HYPOTHALAMUS**

B **SECRETORY NEURONS**

C **ANTERIOR LOBE**

D **POSTERIOR LOBE**

E **OPTIC CHIASMA**

F **BLOOD VESSELS:**

   HYPOTHALAMO-

   HYPOPHYSIAL

   PORTAL SYSTEM

G **HORMONES**

NEUROENDOCRINE SYSTEM: THYROID & PARATHYROID GLANDS

The **thyroid gland** is a bilobed, highly vascular structure, located on the anterolateral aspect of the upper **trachea** and part of the cricoid cartilage. Its secretory cells take up iodine from the plasma and transport it to the colloid where it is used in synthesis of thyroxine. The principal effect of thyroxine is to maintain the proper rate of metabolic activity by stimulating increased oxygen consumption, and is absolutely essential for the proper mental and physical development in the developing fetus and newborn infant.

Hypothyroidism in children (cretinism) is produced by maternal lack of thyroxine during fetal development. Before the advent of iodized salt, cretinism was seen more frequently than it is today. An excessive increase in thyroid activity, hyperthyroidism, is characterized by abnormally high metabolic rates, extreme nervousness and, in some cases, protrusion of the eyeballs (exophthalmos). In those people who lack proper levels of iodine, the level of thyroxine in the plasma is decreased. Consequently, the levels of TSH are increased and the thyroid is stimulated to try to synthesize thyroxine; but without iodine it cannot—all it can do is become hypertrophic and hyperplastic (like trying to go up a downward moving escalator—you put out a lot of energy but you can't go anywhere!). Such is an iodine deficiency goiter.

PARATHYROID GLANDS

These are four, small, peanut-sized structures embedded in the capsule of the thyroid gland on its posterior aspect. Their significance was grasped when it was known that the thyroid itself was not essential to life and yet patients died following total thyroidectomy. The secretion of these bodies is called **parathormone** or simply **parathyroid hormone**.

Parathormone is known to regulate the blood level of ionized calcium, drawing upon bone for a source if the calcium level drops. Decreased parathyroid activity yields the condition called tetany or muscle spasm, resulting in death due to laryngeal constriction if calcium chloride injections are not instituted quickly. Increased parathyroid activity draws excessive calcium out of the bone, leaving a highly porous (cystic) bone that is easily fractured (osteitis fibrosa cystica).

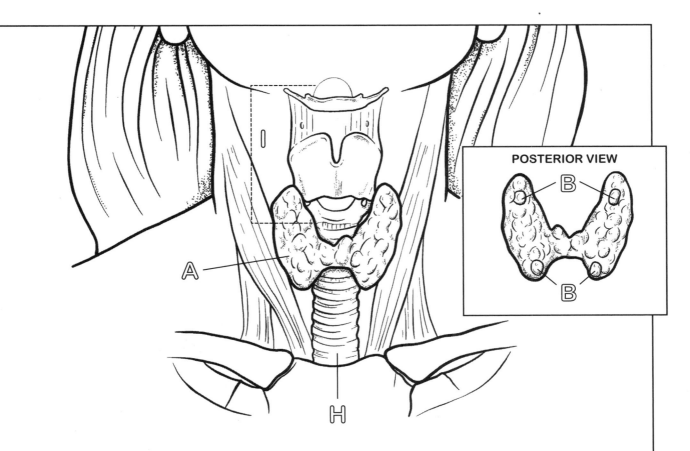

POSTERIOR VIEW

B

B

SECTION OF THYROID GLAND

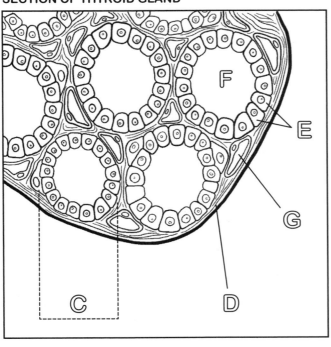

A THYROID GLAND
B PARATHYROID
  GLANDS
C THYROID FOLLICLE
D CONNECTIVE
  TISSUE CAPSULE
E FOLLICULAR CELLS
F COLLOID
G CAPILLARY
H TRACHEA
I LARYNX

REGIONAL ORGANIZATION OF THE HEAD AND NECK
THE SCALP, TEMPORAL & INFRATEMPORAL REGIONS

In the classical tradition, the region of the head is divided up into subregions of the cranium and the face. We've completed our study of the cranial region with the exception of the scalp, the temporal region, and the infratemporal region.

SCALP

The multiple-layered tissue covering the calvaria—often (but sometimes not) replete with hair—is called the scalp. It is quite characteristic of the scalp to bleed profusely when lacerated. The scalp consists of five layers; from superficial to deep, these are:

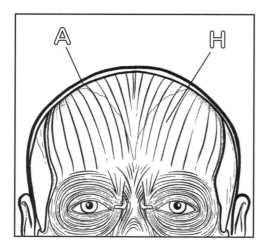

- **Skin**
- **Subcutaneous fascia**
- **Aponeurotic layer**
- **Loose connective tissue**
- **Pericranium**

The skin is thin and incorporates many hair follicles. The **subcutaneous fascia** is a vascular tissue compartmentalized by strong, fibrous strands. Thus when the scalp is lacerated, the blood vessels—strongly enveloped and bound by these strands—cannot contract when cut, and tend to bleed until pressure is applied. The **aponeurosis** of the scalp includes the **occipitofrontalis muscle**. This muscle comes into play whenever the eyebrows are to be raised as in an expression of surprise or horror.

THE TEMPORAL REGION

The temporal region is an area at the side of the skull anterior to and slightly above the ear flap covering the squamous part of the temporal bone. Its lower border is at the level of the zygomatic arch. The principal occupant of this region is the **temporalis muscle** (of mastication), its fascia, and its vessels and nerves, which reach it from the infratemporal fossa.

THE INFRATEMPORAL REGION

This is a well-filled fossa just below the temporal region, bordered externally by the ramus of the mandible and medially by the lateral pterygoid plate. The infratemporal fossa owes its depth in part to the convex zygomatic arch. Some of the contents of this region include:

- Three muscles of mastication, part of the **temporalis** and the lateral and medial pterygoids, and part of the masseter.
- The **maxillary artery** and its branches are a source of blood to the dura, mastication muscles, the gums and teeth of both jaws, the palate, the nasopharynx, and the walls of the nasal cavity.
- The mandibular or third part of the trigeminal nerve (V).
- The chorda tympani nerve innervating the submandibular and sublingual salivary glands and transmitting taste information to the brain from the tongue.
- The pterygoid venous plexus, one of the two important lateral routes of venous return to the brain.

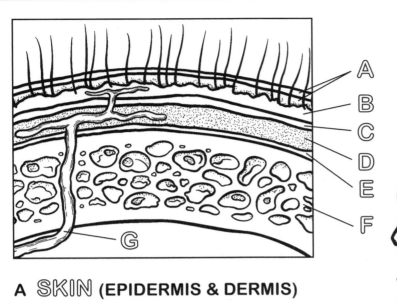

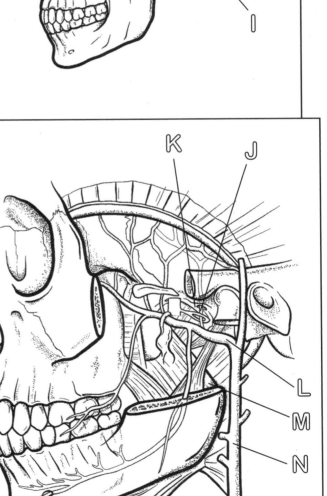

A SKIN (EPIDERMIS & DERMIS)

B SUBCUTANEOUS FASCIA

C APONONEUROSIS

D LOOSE CONNECTIVE TISSUE

E PERICRANIUM

F CRANIUM

G BLOOD VESSEL

H OCCIPITOFRONTALIS MUSCLE

I TEMPORALIS MUSCLE

J MIDDLE MENINGEAL ARTERY

K CHORDA TYMPANI NERVE

L MAXILLIARY ARTERY

M PTERYGOID-MEDIAL MUSCLE

N EXTERNAL CAROTID

REGIONAL ORGANIZATION ACCORDING TO STRUCTURE
MUSCLES OF
FACIAL EXPRESSION

We have yet to consider the muscles of the facial region.

As you make facial expressions note that most of the movements of the face occur with the lips, the nose, and the skin around the eyes. As you may suspect, the majority of facial muscles insert at three places, and you are right. The muscle that allows you to open and close your eyes is the **orbicularis oculi** and arises from the surrounding bone to insert on the skin. The muscles that allow you to twitch or wiggle your nose are called **procerus**, **dilator**, and **compressor nares**.

Most of the muscles of the face are related to the mouth. Note that you can elevate the lips (laughing), depress them (pouting, grief), pucker them (kissing, whistling), retract the angles of your mouth (chagrin or disdain), and just generally create a myriad of expressions through the subtle employment of discrete muscle groups in the face. These muscles are innervated by the fifth cranial nerve.

Muscles that involve the mouth (oris) and lips (labia) are so named:

• **Orbicularis oris**: muscle of the lips that inserts into the orbicularis sphincter of the mouth.
• **Levator anguli oris, levator labii superioris, zygomaticus**: elevators of the lips.
• **Depressor anguli oris, depressor labii inferioris, mentalis**: depressors of the lips.
• **Buccinator**: the "trumpeter's" muscle, a muscle set deep in the cheek, employed for whistling, sucking, and playing musical wind instruments.

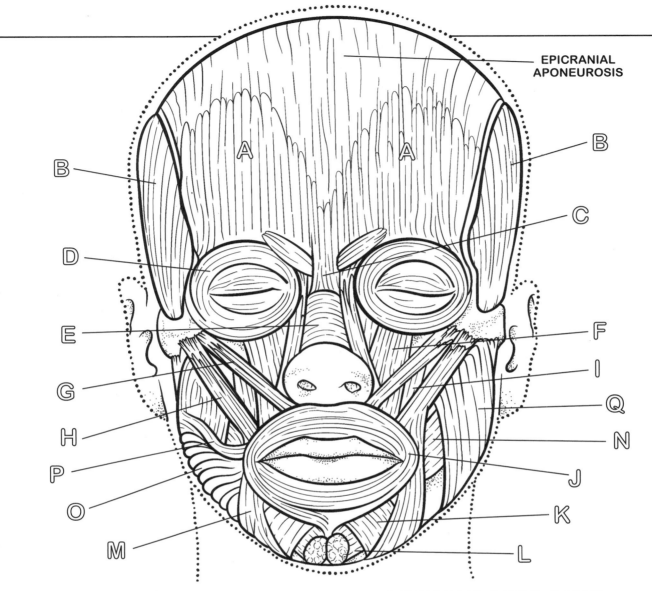

EPICRANIAL APONEUROSIS

A OCCIPITOFRONTALIS	**J** ORBICULARIS ORIS
B TEMPORALIS	**K** DEPRESSOR LABII
C PROCERUS	INFERIORIS
D ORBICULARIS OCULI	**L** MENTALIS **(CUT)**
E NASALIS	**M** DEPRESSOR ANGULI
F LEVATOR LABII	ORIS
SUPERIORIS	**N** BUCCINATOR
G ZYGOMATICUS MINOR	**O** PLATYSMA
H ZYGOMATICUS MAJOR	**P** RISORIUS
I LEVATOR ANGULI ORIS	**Q** MASSETER

REGIONS OF THE NECK: ANTERIOR REGION

The neck is nicely divided into anterior, lateral, and posterior regions.

Anteriorly, the midline of the neck and the two anterior borders of the sternocleidomastoid muscles create a pair of triangles, the **anterior triangles**.

The **sternocleidomastoid** (cleido = clavicle) rotates the head to the opposite side while tilting it downward. Both (left and right) muscles flex the head and neck. The floor of the anterior triangle is made up of the **supra-** and **infrahyoid muscle groups**. These work together in mooring down the hyoid bone and allowing the latter to function at a base for tongue movements. The suprahyoid group are innervated by both cranial and cervical nerves; the infrahyoid group by cervical nerves exclusively.

The roof of the triangles is the most superficial muscle of the neck, the thin platysma, used to tighten the upper neck during shaving. It attaches to the mandible above and skin around the clavicle below.

The contents of the anterior cervical region include:
• The **common, internal,** and **external carotid arteries**.
• The **internal jugular vein**.
• The **vagus nerve** (X). These three occupy a common sheath, bordered anteriorly by the sternocleidomastoid muscle, for the most part.
• The many branches of the external carotid arteries and tributaries of the jugular veins, and the branches of the **hypoglossal** and upper cervical nerves splay out within the triangle.
• The thyroid gland and the embedded parathyroid glands.
• The larynx, clothed in the infrahyoid muscles.
• The laryngopharynx immediately posterior to the larynx.
• Deep muscles of the neck including those that stabilize the first and second ribs during respiration (scalenes) and those that flex the head and neck, lying in front of the vertebrae (prevertebral muscles).

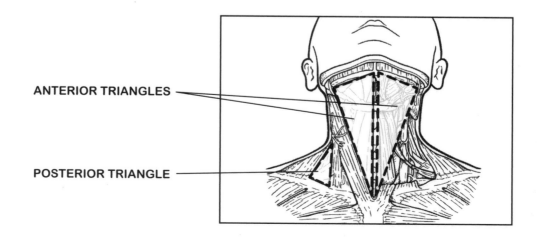

ANTERIOR TRIANGLES

POSTERIOR TRIANGLE

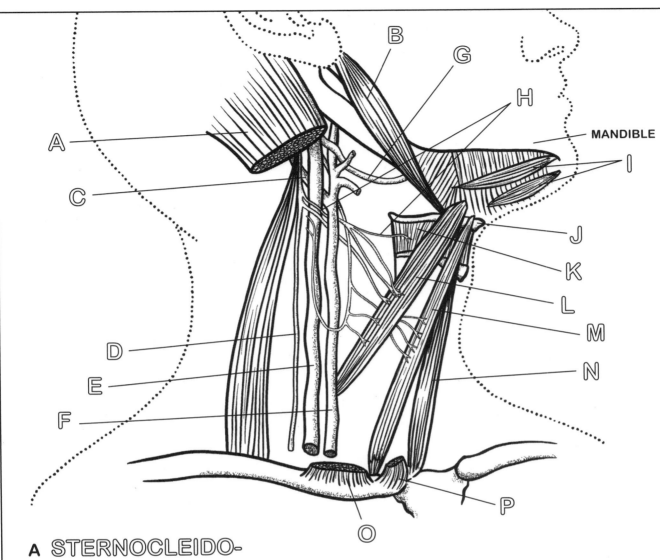

MANDIBLE

A STERNOCLEIDO-
MASTOID MUSCLE

B STYLOHYOID

C INTERNAL CAROTID
ARTERY

D VAGUS NERVE

E INTERNAL JUGULAR
VEIN

F COMMON CAROTID
ARTERY

G HYPOGLOSSAL NERVE

H ANSA CERVICALIS
(CERVICAL NERVES 1–4)

I SUPRAHYOID GROUP

J HYOID BONE

K THYROHYOID

L OMOHYOID

M STERNOHYOID

N STERNOTHYROID

O CLAVICULAR HEAD **(CUT)**

P STERNAL HEAD **(CUT)**

REGIONS OF THE NECK: LATERAL REGION

The relationship of the superficial muscle of the back, the **trapezius**, with the **clavicle** and the **sternocleidomastoid** muscle form a triangle which is referred to as the **posterior triangle**. This supraclavicular hollow (triangle) incorporates several vascular and neural structures traversing this area, including:

• The **external jugular vein** (frequently visible, with visible valves).
• The accessory nerve (XI) to trapezius.
• The deeper **brachial plexus** en route to the axilla from the spinal cord.
• **Cutaneous nerves** to surrounding muscle.
• Arteries and veins en route to the superficial back from the subclavian artery.
• **Cervical lymph nodes**.

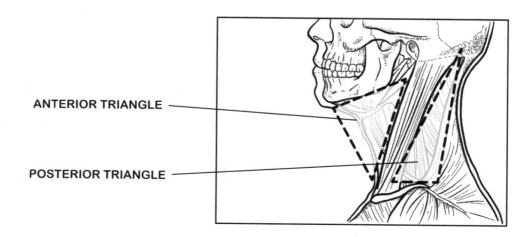

ANTERIOR TRIANGLE

POSTERIOR TRIANGLE

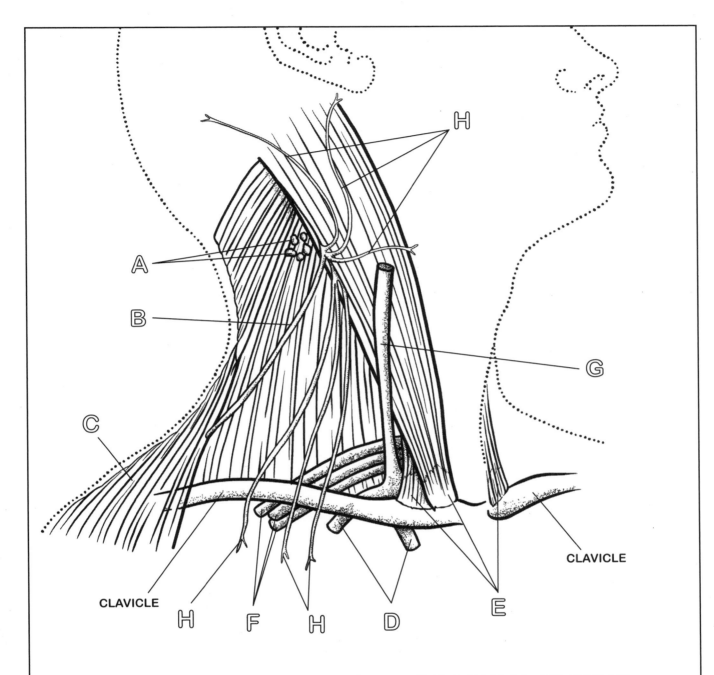

A CERVICAL LYMPH NODES
B ACCESSORY NERVE
C TRAPEZIUS
D SUBCLAVIAN VESSELS
E CLAVICULAR & STERNAL HEADS OF STERNOCLEIDOMASTOID

F BRACHIAL PLEXUS
G EXTERNAL JUGULAR VEIN
H CUTANEOUS NERVES OF THE CERVICAL PLEXUS

REGIONS OF THE NECK: POSTERIOR REGION

This region is stuffed with layer after layer of muscle and lies within a boundary drawn from C7 to the occiput above, and from the lateral border of the **trapezius** on one side to the lateral border on the other side. The deep portion of this region just under the occipital bone (suboccipital region) includes a rather fascinating triangular arrangement of muscle bundles involved in rotating, flexing, and extending the head on the cervical vertebrae such that "yes" and "no" motions are possible. All are supplied by dorsal rami of upper cervical nerves. A probe pushed through the center of the muscular triangle on each side will pass through the vertebral artery en route to the brain from the subclavian artery.

ARTERIES AND VEINS

The sources of blood to the neck and head are the common carotid arteries arising directly (left) and indirectly (right) off the arch of the aorta. These arteries bifurcate into internal and external branches at about the angle of the mandible.

The brain, face, and neck are all drained by the internal and external jugular vein and drain into the right side of the heart via the brachiocephalic vein and superior vena cava. The internal jugular vein is assisted by the pterygoid and the ophthalmic venous plexuses in draining the brain and face. The location of the tributaries of the jugular veins is extremely variable.

INNERVATION OF THE HEAD AND NECK

The scheme of head and neck innervation may be seen in the adjacent illustration. Sympathetic innervation is derived from upper thoracic preganglionic neurons, which synapse in the superior, middle, and inferior **cervical ganglia**. The postganglionic fibers reach their destinations by "hitching a ride" on one of the local arteries, e.g., the internal carotid artery to the orbit. These fibers are principally vasoconstrictors in effect and also dilate the pupil of the eye.

Parasympathetic innervation of the head and neck is derived from cranial nuclei associated with the third, seventh, ninth, and tenth cranial nerves. These fibers innervate the glands of the head and muscles of the eye.

The general somatic innervation of the head and neck includes the cutaneous fibers serving the head and the fibers supplying the muscles of mastication which are largely branches of the trigeminal cranial nerve (V). The back of the head and the entire neck are supplied by branches of the first to fourth **cervical nerves**.

LYMPHATIC DRAINAGE OF THE HEAD AND NECK

All lymphatic vessels of the head and neck ultimately drain into a chain of lymph nodes under cover of the sternocleidomastoid muscle. These are the deep cervical lymph nodes and receive the flow of such nodes as the parotid, facial, and submaxillary.

Note: The left side of the ilustration shows a deeper region than the right side.

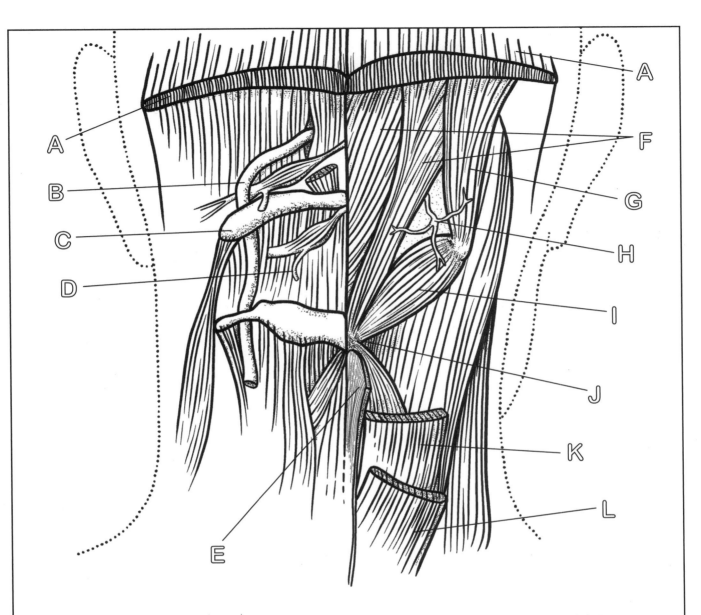

A TRAPEZIUS **(CUT)**

B VERTEBRAL ARTERY

C TRANSVERSE PROCESS
OF C1

D SECOND CERVICAL
NERVE & GANGLION

E LIGAMENTUM NUCHAE

F RECTIS CAPITA
POSTERIOR **(MAJOR & MINOR)**

G SUPERIOR OBLIQUE

H FIRST CERVICAL NERVE

I INTERIOR OBLIQUE

J SPINE OF C2

K SEMISPINALIS **(CUT)**

L SPLENIUS **(CUT)**

THE CENTRAL NERVOUS SYSTEM: BRAIN & SPINAL CORD

Encased in the skull is the **brain**, the central organ of the human nervous system. The brain and **spinal cord** together are the **central nervous system** (CNS), which receives information from all over the body—with the help of the peripheral nervous system (PNS)—and then coordinates activity to produce appropriate body responses. The brain is the command center, and the spinal cord is a column of nerve tissue that carries signals between the brain and body.

As we learned on page 52, the fundamental unit of the nervous system is the **neuron**, which consists of a cell body and processes. In the CNS, there are billions of neurons! Collections of these cell bodies are called **nuclei**; and collections of nerve cell processes, or axons, are called **tracts**. Tracts transmit the electrical signals nuclei use to communicate. Their axons are encased in whitish **myelin** to better conduct the electrical signals, and as such, myelin-dense tracts in the brain and spinal cord are referred to as **white matter**. Nuclei in the related processes and their supporting glial cells are generally not myelinated and so are grayish— known as **gray matter**.

White and gray matter, mechanically fastened and metabolically supported by nonneural connective tissues found only in the brain and spinal cord, make up the entire central nervous system. Gray matter is mainly found on the brain's outermost surface, and white matter is deeper within the brain. Conversely, gray matter is at the core of the spinal cord, with white matter as the outer layer.

Remember that the human nervous system is a vast communications network. Incoming data from receptors and sensory neurons of the PNS percolate through the myriad neurons of the CNS and, by the coordinated efforts of its nuclei and centers, appropriate responses are issued.

Let us learn more about the complex central organ of the human nervous system.

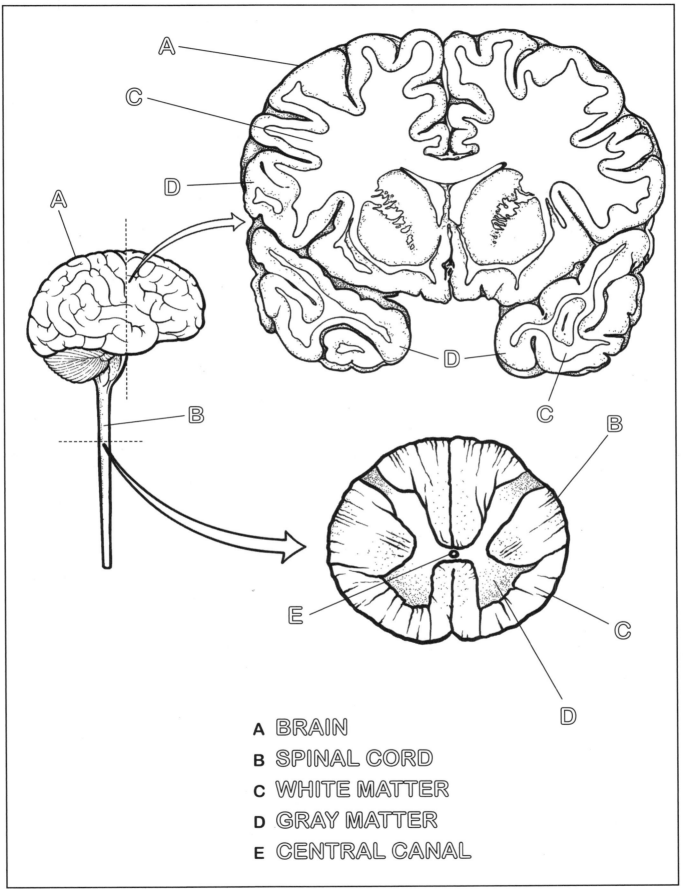

A BRAIN
B SPINAL CORD
C WHITE MATTER
D GRAY MATTER
E CENTRAL CANAL

THE BRAIN:
EARLY GROWTH STAGES

The key to understanding this most essential organ lies in appreciating its humble beginnings. At twenty-five days of age, the embryo is in the midst of extensive external and internal development. In the posterior midline region just deep to the skin, a hollow tube of embryonic nerve tissue has taken form. This tube will undergo considerable growth, especially in the head end where the brain is to develop. The caudal two thirds of this hollow neural tube will become the **spinal cord**. Soon after the neural tube has taken shape, three regions of the developing brain can be distinguished:

• The **forebrain**, which grows into two somewhat hollow hemispheric vesicles, one on each side, creating the **telencephalon** (end brain). The remaining portion of the forebrain is squeezed between the hemispheres of the telencephalon, and is appropriately renamed the **diencephalon** (between brain).

• The **midbrain** largely retains its external shape and **mesencephalon** (middle brain) remains its name.

• The **hindbrain** will undergo considerable rearrangement. The upper part develops characteristic outpocketings anteriorly and posteriorly. This region is named the **metencephalon** (after brain). The lower hindbrain, as one moves inferiorly, looks increasingly like the spinal cord and is suitably named the **myelencephalon** (spinal brain).

In subsequent months of development, the most spectacular growth continues in the telencephalon. The diencephalon, hidden between the hemispheres, enlarges. In the metencephalon, the anterior swellings project posteriorly to join with the highly convoluted posterior prominence, with relatively little external change except to increase in mass. The same may be said for the spinal cord. It should be understood that highly complex internal development goes on despite minor external changes. The important derivatives of these five basic regions are as follows:

1. Telencephalon	2. Diencephalon	3. Mesencephalon

a. Cerebral cortex and related white matter
     b. Basal (subcortical) nuclei
	a. Thalamus
	     b. Hypothalamus
		a. Cerebral peduncles
		     b. Superior colliculi
		     c. Inferior colliculi

4. Metencephalon	5. Myelencephalon
a. Pons **b. Cerebellum**	**a. Medulla**

The cavity within the developing neural tube changes along with each brain region. The cavity of the forebrain undergoes considerable dilation, particularly in the telencephalon where it develops into bilateral lateral ventricles (first and second). It thins into a midline, flat third ventricle in the diencephalon, retaining communications with the lateral ventricles via two interventricular foramina. Continuing caudally from the third ventricle, the cavity narrows into a tube-like cerebral aqueduct within the substance of the midbrain.

Going into the metencephalon, the cavity flares out to become the fourth ventricle. It narrows within the myelencephalon to merge into the central canal of the spinal cord.

The fluid of the ventricles is called **cerebrospinal fluid** (CSF), a plasma-like material, which is secreted from the **choroid plexus**—an epithelial tissue heavily infiltrated with capillaries on the roof of the lateral, third, and fourth ventricles.

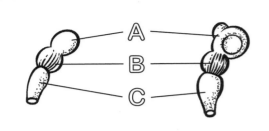

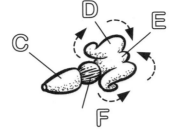

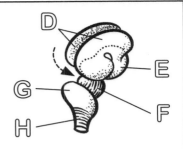

22-DAY EMBRYO　　**28-DAY EMBRYO**　　**POSTERIOR VIEW**　　**LATERAL VIEW**

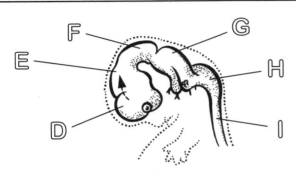

7-WEEK EMBRYO

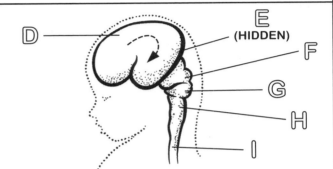

(HIDDEN)

11-WEEK FETUS

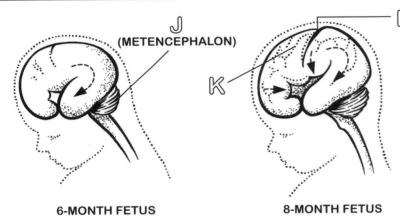

J (METENCEPHALON)

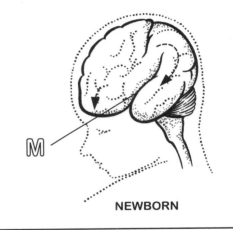

L

6-MONTH FETUS　　**8-MONTH FETUS**　　**NEWBORN**

A FOREBRAIN

B MIDBRAIN

C HINDBRAIN

D TELENCEPHALON

E DIENCEPHALON

F MESENCEPHALON

G METENCEPHALON

H MYELENCEPHALON

I SPINAL CORD

J CEREBELLUM

K GYRUS

L CENTRAL SULCUS

M LATERAL CEREBRAL
FISSURE

THE BRAIN: CEREBRAL HEMISPHERES

The brain is oriented into five regions. It is important to emphasize the continuity of the structure in the CNS and to understand the interregional continuity when considering tracts and pathways of the CNS and the nuclei with which they relate.

The CNS is organized as follows:

• **Cerebral hemispheres**
• **Brainstem**
• **Cerebellum**
• **Spinal cord**

Let us explore each.

CEREBRAL HEMISPHERES

The **cerebral hemispheres** take up the largest share of the brain mass—those cauliflower-like structures bulging out above and over the brainstem like a great flower projecting from its stem. The hemispheres are separated by a deep longitudinal fissure and attached by a flat white band of interconnecting tracts. They are further characterized by a lateral fissure separating the temporal lobe from the more central part of the cerebral hemispheres.

Over the next pages we will learn about the cerebral hemispheres and their structures.

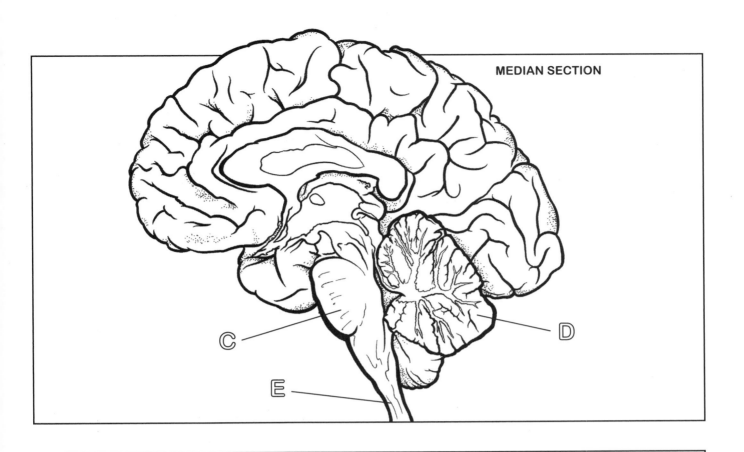

C

D

E

SUPERIOR VIEW   **INFERIOR VIEW**

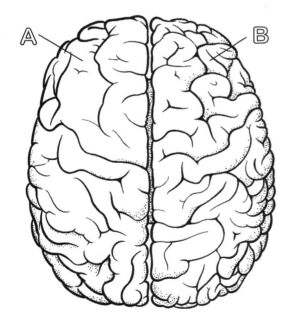

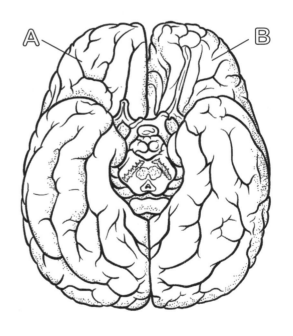

A

B

A

B

CEREBRAL HEMISHPERES:

A LEFT HEMISPHERE

B RIGHT HEMISPHERE

C BRAINSTEM

D CEREBELLUM

E SPINAL CORD

CEREBRAL HEMISPHERES
BRAIN FUNCTION
ACCORDING TO LOBES

Each cerebral hemisphere of your brain contains four **lobes**, with deep furrows and round ridges. In anatomical parlance, such a furrow is called a **sulcus** (pl., sulci); a ridge is called a **gyrus** (pl., gyri). These sulci and gyri increase the surface area of the brain considerably. These sulci and fissures of the cortex are employed by the neuroanatomist to divide the cortex and underlying white matter into lobes of which:

• Four take the names of the bones shielding them: **frontal**, **parietal**, **temporal**, and **occipital**.
• One is set deep into the lateral fissure and is not visible at the surface: the **insula**.
• One is not a distinct anatomical entity but a composite of several structures from different areas: the **limbic lobe**.

Cortical activities have been assigned to lobes. Within a lobe, precise areas for specific function do not exist. Functions overlap from area to area and from lobe to lobe.

The **frontal lobe** is generally concerned with voluntary and reflex motor activity—that is, these activities are largely initiated here. It is believed that the "higher functions" of humans, such as abstract reasoning, learning, and intelligence, are a product of frontal activity. Memory takes place here as it does in most places of the cortex. Expression through speech may be related to this region.

The **parietal lobe** incorporates the receiving areas for such somesthetic sensations as pain, temperature, pressure, and touch. Only when these impulses reach the parietal lobe does one become aware of these sensations. Recognition of your body's position in space (proprioception) is also registered here. The parietal lobe is also related to speech and the interpretation of language. Like all other cerebral lobes, the parietal lobe has abundant connections to other parts of the cortex, the brainstem, and the cerebellum.

The **temporal lobe** receives impulses related to hearing and integrates them with other sensory input and memory as well. Localization and awareness of sound is produced here. Bilateral removal of the auditory cortex causes deafness. Because the medial and inferior aspects of the temporal lobe are anatomically and functionally related to the limbic and frontal lobes, certain aspects of behavior and emotional expression as well as memory patterns are associated with the temporal lobe. Comprehension of language—both spoken and written—is also made possible here.

The **occipital lobe** is almost entirely devoted to the reception of visual input and associating this input with memory and the data of other areas of the cortex. Destruction of this area results in a spectrum of defects, the worst being complete blindness.

The **limbic lobe** and functionally related areas (collectively called the **limbic system**) have complex tasks not totally understood. Generally the limbic lobe is associated with activities relating to self-preservation and preservation of the species, such as eating, fighting, fear, flight to safety, sexual behavior, and parental care. The limbic lobe plays an important role in making our response to stimuli subjective (based on feeling) rather than objective (based on intellect). Bias and prejudice may well be related to limbic activity. The limbic lobe has extensive connections with the hypothalamus of the brainstem.

In summary, it is in the cerebral cortex that sensory awareness takes place and that voluntary and reflex motor activity is initiated. In addition, it is here that the higher faculties of humans are found. Our increasing ingenuity throughout evolutionary history is partly a reflection of increasing cortical complexity—a complexity based on expanding numbers of neurons and, as a consequence, their interconnections.

Note: The different lobes of the brain are separated by dotted lines on the illustration.
Use different colors for each area within the dotted lines. Have fun choosing your colors!

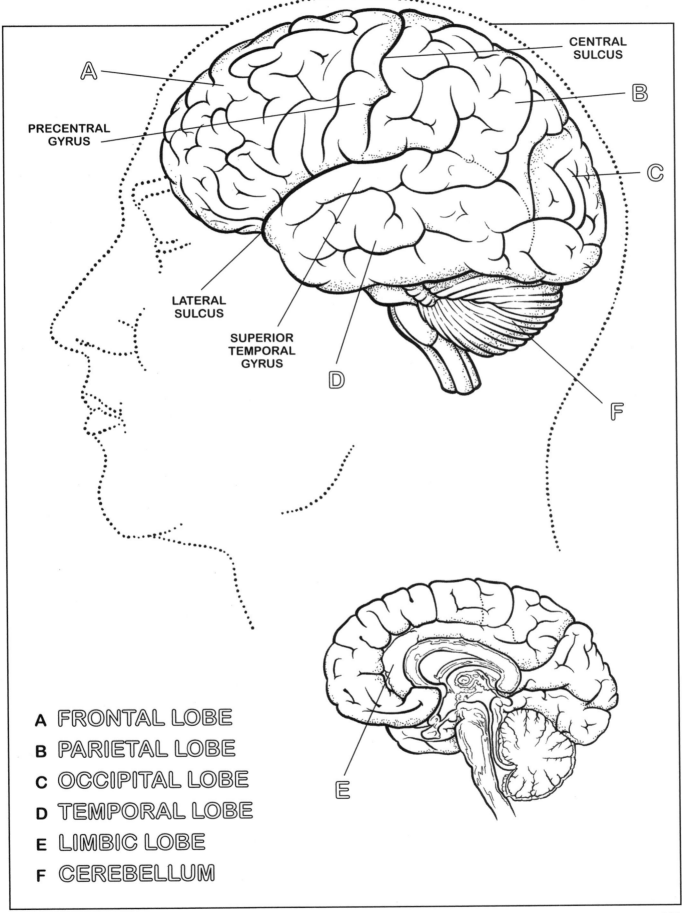

CENTRAL
SULCUS

PRECENTRAL
GYRUS

LATERAL
SULCUS

SUPERIOR
TEMPORAL
GYRUS

A FRONTAL LOBE
B PARIETAL LOBE
C OCCIPITAL LOBE
D TEMPORAL LOBE
E LIMBIC LOBE
F CEREBELLUM

CEREBRAL HEMISPHERES
CEREBRAL CORTEX: GRAY MATTER

The brain plays a central role in the control of most of our bodily functions: movements, sensations, awareness, thoughts, speech, and memory—and **gray matter** enables us to do it. The **cerebral cortex** is the external layer and surface structure of the brain, consisting of gray matter: neural cell bodies, axon terminals, and dendrites (as well as all nerve synapses). It is named for its pinkish-gray color. These neurons of the cerebral cortex generally occupy one of the following areas:

SENSORY AREAS OF THE CORTEX

Neurons in the various sensory areas of the cortex are concerned with the final receipt of ascending impulses from lower centers in the brain and spinal cord. It is in these areas that one is made aware of a sensory experience. More specifically, in these areas one is able to discriminate among the various sensations the body is equipped to perceive. The following are the principal sensory areas of the cortex:

• **Sensory** impulses related to conscious perception of pain, temperature, touch, pressure, and muscle sense reach the postcentral gyrus of the **parietal lobe**.
• **Visual** impulses arrive at the calcarine fissure of the **occipital lobe**.
• **Auditory** impulses related to the conscious perception of sound terminate in a portion of the superior temporal gyrus of the **temporal lobe**.
• The **olfactory** system, or sense of smell, is believed to make contact with a complex area relating to the limbic system at the medial and inferior aspects of the **temporal lobe**.

HIGHER COGNITIVE FUNCTION AREAS OF THE CORTEX

The **frontal lobe** is important for **higher mental functions** such as memory, emotions, impulse control, organization, problem solving, and social interaction.

MOTOR AREAS OF THE CORTEX

The principal **motor** area is located largely in the precentral gyrus of the **frontal lobe**. The neurons in this area initiate willful, skilled movement on the opposite side of the body. These impulses are largely transmitted down a specific tract called the corticospinal tract or pyramidal pathway to the motor neurons of the spinal cord. There are also parts of the frontal lobe whose cortical neurons initiate unskilled, reflex postural movements. The axons of these neurons are part of the corticospinal tract.

ASSOCIATION AREAS OF THE CORTEX

All areas of the cortex not specified as "motor" or "sensory" are generally labeled **association areas**. They integrate sensory input of different kinds with memory (also stored in association areas), permitting complex perceptions and emotional expressions. For instance, you see a picture of a car and recall the sound of a powerful engine; your friend sees the same picture and recalls the smell of a new car's interior; another friend sees the picture and immediately feels anxious because not long ago he was in an automobile accident.

Association areas also provide the mechanism for complex mental activities such as reasoning and thinking in abstract terms. In addition, association areas enable complex motor activities such as talking or walking. Ultimately, association areas integrate sensory input to provide motor output appropriate for the occasion.

Note: The different funtional areas of the brain are separated by dotted lines on the illustration. Use different colors for each area within the dotted lines. Have fun choosing your colors!

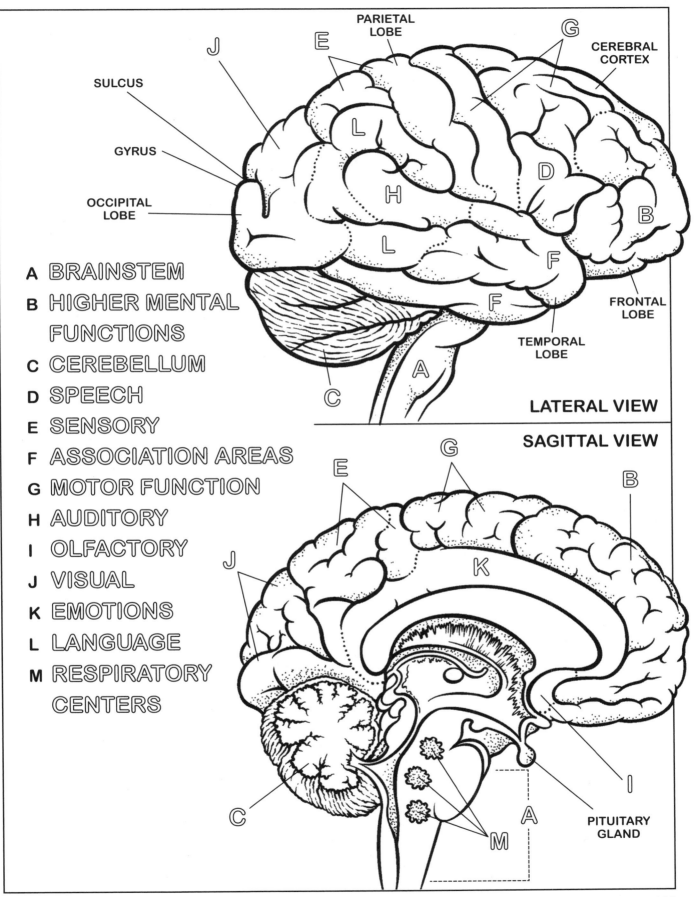

PARIETAL LOBE

CEREBRAL CORTEX

SULCUS

GYRUS

OCCIPITAL LOBE

FRONTAL LOBE

TEMPORAL LOBE

LATERAL VIEW

A BRAINSTEM
B HIGHER MENTAL FUNCTIONS
C CEREBELLUM
D SPEECH
E SENSORY
F ASSOCIATION AREAS
G MOTOR FUNCTION
H AUDITORY
I OLFACTORY
J VISUAL
K EMOTIONS
L LANGUAGE
M RESPIRATORY CENTERS

SAGITTAL VIEW

PITUITARY GLAND

CEREBRAL HEMISPHERES
WHITE MATTER

As you may recall, the fundamental unit of the nervous system is the neuron, which consists of a cell body and processes (see page 52). Collections of cell bodies are called **nuclei** and collections of nerve cell processes, or axons, are called **tracts**. Found in the deeper, subcortical tissues of the brain, nuclei tracts have axons that are whitish in color and are thus referred to as **white matter**: millions of bundles of axons that connect neurons from different brain regions to *functional circuits*. The white color derives from the myelin coating the axons. Myelin acts as insulation and is essential for high-speed transmission of nerve cell electrical impulses. In general, these collections of myelinated nerve processes take one of three courses:

COMMISSURES
Running from one hemisphere to another are **commissures**. The largest of these is the **corpus callosum** which runs longitudinally from anterior to posterior and lies deep in the cortex, connecting one lobe or region with another in the same hemisphere through association tracts.

PROJECTION FIBERS
Probably the most spectacular of all fiber systems in the brain, **projection fibers** travel vertically from higher centers to lower centers and vice versa. The chief system is a band of fibers projecting up through the substance of each hemisphere (like the stalks of a flower bouquet) called the **internal capsule**. As it fans out to all regions of the cortex (like the flowers themselves) it gives one the impression of a "radiating crown" and is so named: **corona radiata**. The internal capsule and corona is the primary avenue of fiber communication between the cerebral cortex and lower centers.

BASAL (SUBCORTICAL) NUCLEI
These are groups of neuron cell bodies (nuclei) making up, in part, the basal and medial walls of each hemisphere. They include:

• The **caudate nucleus**, lying between the lateral ventricle and the internal capsule. It is a tail-shaped structure following the contours of the lateral ventricle into the medial portion of the temporal lobe where it terminates as the almond-shaped amygdala (n Latin, it means almond).

• The **lenticular nucleus**, partly encircled by the caudate and separated from it by the internal capsule, is so named because of its "lens" shape. It consists of two nuclei called the **putamen** and **globus pallidus**.

Basal nuclei, in conjunction with lower brainstem nuclei, are principally concerned with reflex movements such as postural adjustments—positioning the body for a golf swing or batting a baseball. Impulses from the basal nuclei probably do not initiate movement; they modify cortex-initiated impulses. In this respect, basal nuclei are believed to play an error control function preventing any movement other than that desired. Destruction of certain basal nuclei results in diseases associated with abnormal movements (dyskinesia).

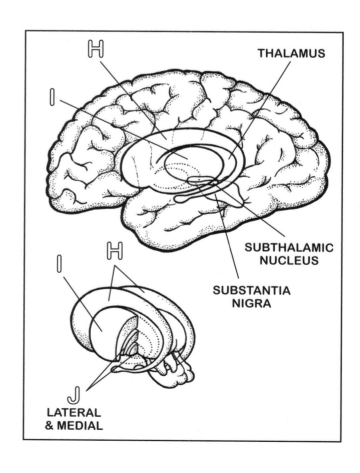

THALAMUS

SUBTHALAMIC NUCLEUS

SUBSTANTIA NIGRA

LATERAL & MEDIAL

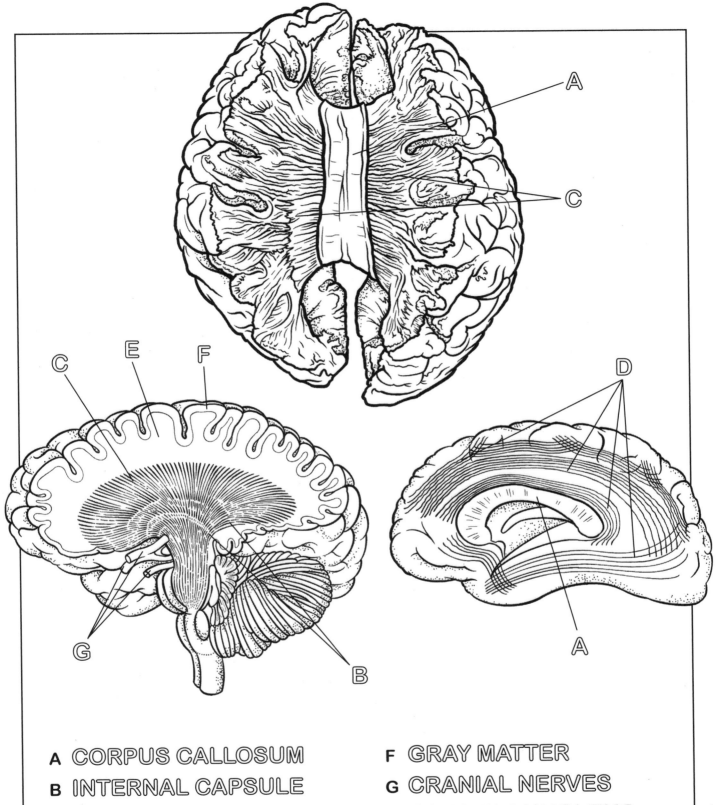

A CORPUS CALLOSUM

B INTERNAL CAPSULE

C CORONA RADIATA

D AXON TRACTS

E WHITE MATTER

F GRAY MATTER

G CRANIAL NERVES

H CAUDATE NUCLEUS

I PUTAMEN

J GLOBUS PALLIDUS

THE BRAINSTEM

The **brainstem** includes the midbrain, pons, medulla, thalamus, and hypothalamus. It makes connections with the cerebral hemispheres by way of the internal capsule and the cerebellum by way of the three paired cerebellar fiber bundles known as **peduncles** (superior, middle, inferior).

Except in the diencephalon, three regions can be identified in most cross sections taken through the brainstem:
• The **tectum**: uppermost part of the midbrain and at the rear of the cerebral aqueduct. The tectum incorporates the superior and inferior colliculi of that region, but is thin and without much significance in the rest of the brainstem.
• The **base**: the anterior aspect of the midbrain, pons, and medulla.
• The **tegmentum**: the region between the tectum and the base. It is separated from the tectum by the cerebral aqueduct and fourth ventricle, and the third and fourth cranial nerves supply some of the extrinsic muscles of the eye.

MIDBRAIN

The midbrain consists of the tectum incorporating the superior and inferior colliculi, the tegmentum, and the cerebral peduncles.

The **colliculi** are four knobs of gray matter, with the superior pair communicating with the optic tracts—while the inferior pair are believed to integrate reflex movements of the head to auditory input, such as seen in one's own response to alarming sounds (firecrackers, horns, etc.).

The reticular formation of the brainstem consists of a diffuse network of neurons that percolate ascending and descending impulses through a myriad of synapses. These impulses have facilitatory or inhibitory influences on almost any aspect of sensory input or motor output, e.g., sleep, wakefulness, awareness, reflexive movements, muscle tone, etc. The distinctive red nucleus of the midbrain is part of the midbrain reticular formation.

The **cerebral peduncles** consist largely of tracts descending from the cerebral cortex (via the internal capsule) to the spinal cord (corticospinal tract) and to intermediate nuclei (corticopontine and corticobulbar tracts).

PONS

The **pons** (in Latin, it means bridge) consists of a deeper tegmentum and a more superficial basilar part of the brainstem that links the medulla oblongata and the thalamus.

MEDULLA

The **medulla** is the continuation of the spinal cord within the skull, forming the lowest part of the brainstem and containing control centers for the heart and lungs. It is clearly separated from the pons by a transverse sulcus and merges indistinctly into the spinal cord at the level of the foramen magnum.

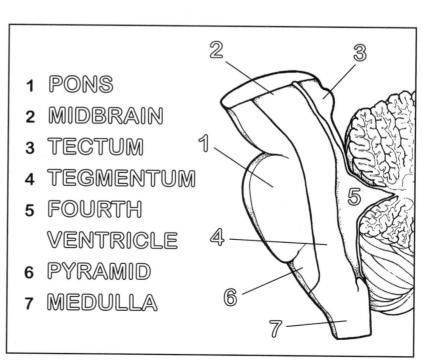

1 PONS
2 MIDBRAIN
3 TECTUM
4 TEGMENTUM
5 FOURTH VENTRICLE
6 PYRAMID
7 MEDULLA

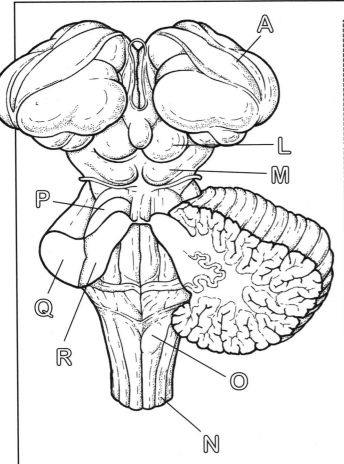

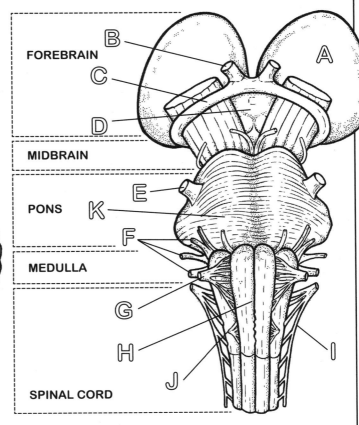

FOREBRAIN

MIDBRAIN

PONS

MEDULLA

SPINAL CORD

A THALAMUS
B OPTIC NERVE
C OPTIC CHIASMA
D FLOOR OF HYPOTHALAMUS
E TRIGEMINAL NERVE
F NERVES
G HYPOGLOSSAL NERVE
H PYRAMID
I ACCESSORY NERVE

J VENTRAL ROOT OF FIRST CERVICAL NERVE
K PONS
L SUPERIOR COLLICULI
M INFERIOR COLLICULI
N SPINAL CORD
O MEDULLA OBLONGATA

CEREBELLAR PEDUNCLES:

P SUPERIOR
Q MIDDLE
R INFERIOR

THE BRAINSTEM
THALAMUS & HYPOTHALAMUS

THALAMUS

The **thalamus** is a paired, football-shaped mass of many nuclei immediately lateral to the third ventricle. It is connected to its fellow on the opposite side by an intermediate mass of gray matter. In very general terms, the thalamus functions as a center: receiving impulses from ascending (sensory), association, and some descending fibers; correlating, integrating, and distributing impulses to appropriate areas of the cerebral cortex, basal nuclei, hypothalamus, cerebellum, and other brainstem nuclei.

HYPOTHALAMUS

The single, unpaired **hypothalamus** is a most compact and complex aggregation of nuclei and related processes lying below and slightly anterior to the thalamus, hugging the lower part of the third ventricle. The circuitry to and from the hypothalamus is extensive—connections being made to nuclei throughout the cerebral hemispheres and brainstem. The functions generally attributed to this area are a memorizer's dream:

• Regulation of sympathetic and parasympathetic activity, working, in part, through the cardiovascular and respiratory centers in the medulla. Any expression of emotion incorporating visceral changes (changes in heart rate, blood pressure, respiratory rate, etc.) involve the hypothalamic nuclei.

• In association with the hypophysis (pituitary), it influences certain endocrine activity, e.g., lactation, ovulation, onset of puberty, the menstrual cycle, water balance, and general growth development.

• It regulates body temperature by influencing mechanisms of heat production and conservation, including vasodilation (enlargement of vessels' lumina), vasoconstriction (narrowing of the lumina), etc.

• It regulates appetite, possibly being signaled by subtle changes in sugar concentration in the plasma and distension of the stomach.

• In association with the reticular system, the hypothalamus influences the sleep-wakefulness mechanism.

EPITHALAMUS

Perhaps better known as the **pineal gland**, this region of the diencephalon seems to play a role in the menstrual cycle in females—the significance of which is still being studied in humans. The pineal overlies the superior colliculi on the posterior aspect of the brainstem. What has been learned in recent years is that this gland produces melatonin and influences the light-dark cycle from the environment and directly affects the sleep/wake cycles for individuals.

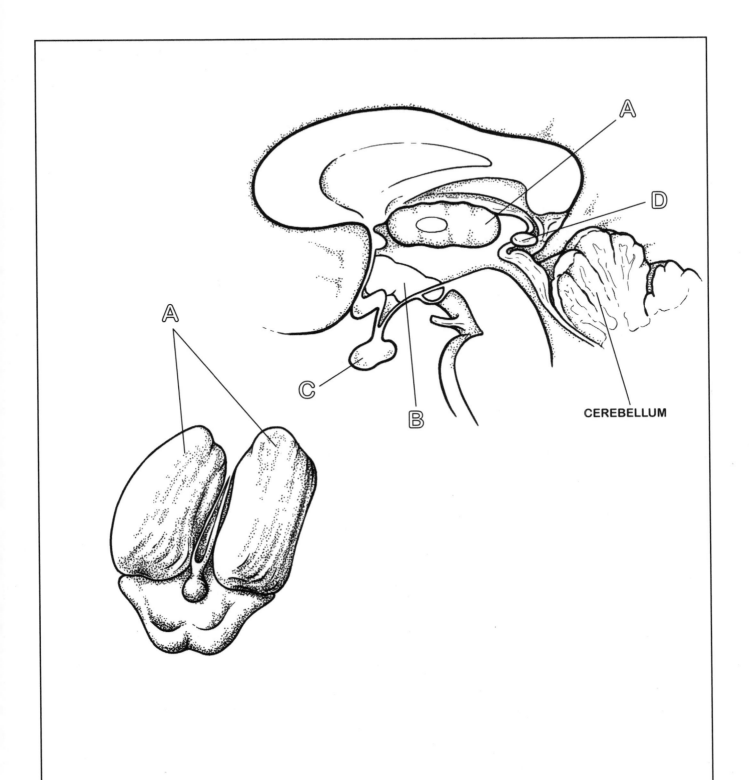

A THALAMUS

B HYPOTHALAMUS

C HYPOPHYSIS (PITUITARY GLAND)

D EPITHALAMUS (PINEAL GLAND)

CEREBELLUM

THE CEREBELLUM

The **cerebellum**, although derived from the metencephalon along with the pons, is not part of the brainstem. Its Latin name, meaning "little brain," refers to the structural similarity to the cerebral hemispheres. The cerebellum is principally responsible for muscular coordination. Based on input from proprioceptors in the muscles and tendons (via the inferior and middle cerebellar peduncles), the cerebellum assures proper muscle tension and the coordinated contraction of several muscle groups to effect a precise movement, such as picking up a pebble.

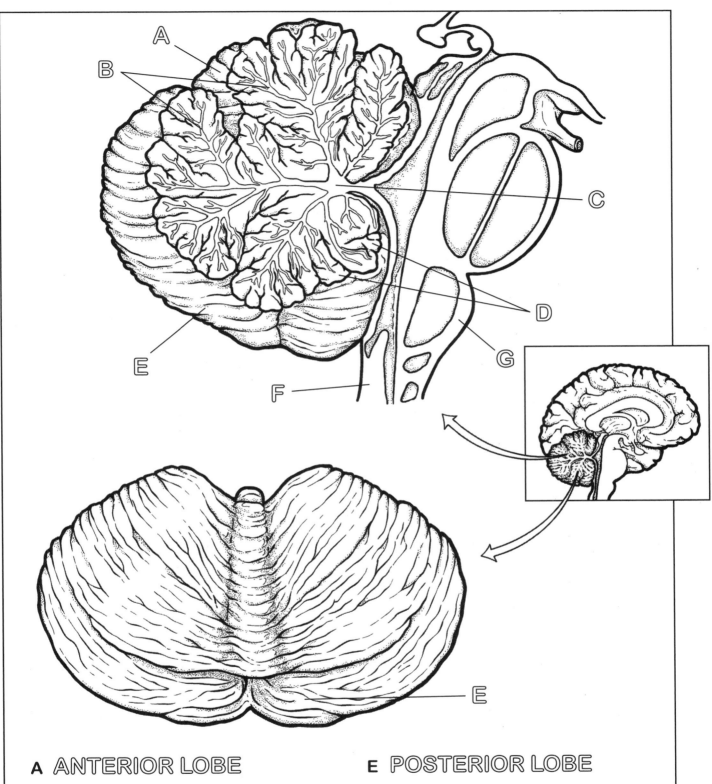

A ANTERIOR LOBE
B ARBOR VITAE
C CEREBELLAR NUCLEUS
D CEREBELLAR CORTEX
E POSTERIOR LOBE
F SPINAL CORD
G MEDULLA OBLONGATA

THE SPINAL CORD

The brain has been described as the grand integrator, coordinator, and modifier of sensory information and motor response. The **spinal cord** functions to extend the authority of the brain to the body proper and provides a means by which the body's peripheral structures can get quick access to the brain for awareness and decisions. Stub your toe and see how long it takes for cortical awareness!

The spinal cord takes up the lower two thirds of the central nervous system, beginning at the level of the foramen magnum as the distal continuation of the **medulla**. Locked within the framework of the vertebral column, the spinal cord ends at the level of the **second lumbar vertebra**.

The spinal cord is generally responsible for:
• Providing the connections between incoming and outgoing reflexes to the limbs.
• Routing input from sensory (peripheral) nerves to appropriate motor cells in the cord (for the simplest reflexes), to cells of the brainstem (for all but the simplest reflex acts), and to the cerebral cortex (for awareness).
• Routing of all impulses descending below the medulla on to the motor neurons, which generate the actual motor impulses to the muscles and glands of the body.

There are five regions of the spinal cord that correspond to the vertebrae enclosing them, i.e. cervical, thoracic, lumbar, sacral, and so on. The cord gives off the roots of the spinal nerves. The cord gives off the roots of the spinal nerves. The **dorsal** and **ventral** roots merge to form a single unit (the **spinal nerve**, see following page) as they pass out of the vertebral canal through the **intervertebral foramina** (spaces between adjacent vertebrae). In so doing, they take with them their covering of dura mater, which is continuous with the neurilemma of the peripheral nerves. Each bilateral pair of spinal nerves is numbered according to the region of the spinal cord from which it arises. The first seven cervical nerves emerge above their respective vertebrae—that is, C1 nerve emerges above C1 vertebrae. The eighth cervical nerve and all subsequent spinal nerves emerge below their respective vertebrae—that is, C8 nerve emerges below C8 vertebrae.

Enlargements of the cord are visible at the lower cervical region and upper lumbar regions of the vertebral column. These **cervical** and **lumbosacral enlargements** represent those portions of the cord contributing to the brachial and lumbosacral plexuses.

The roots coming off the cord are directed progressively downward as one moves along the cord caudally. The cord does not fill the extent of the vertebral canal.

The **conus medullaris** is the terminal end of the spinal cord, surrounded by streams of spinal nerve roots (*cauda equina*, "horse's tail") at the level of the L1–2. All of these structures are within the **dural sac**, which ends at S2.

The cord is tied down to the coccygeal vertebrae by a thin thread (**filum terminale**) of meningeal tissue extending from the conus terminale.

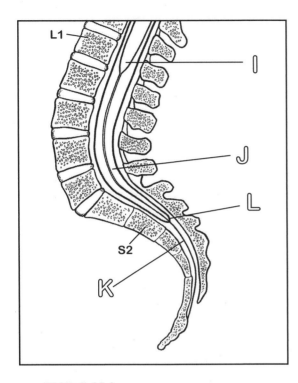

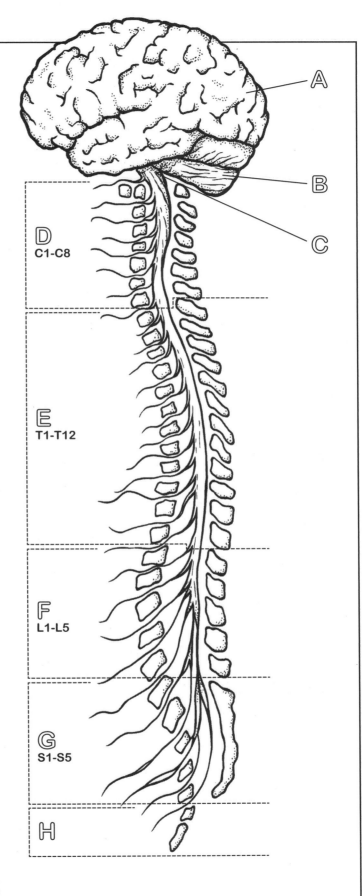

L1

I

J

L

S2

K

A BRAIN
B CEREBELLUM
C MEDULLA
D CERVICAL
E THORACIC
F LUMBAR
G SACRAL
H COCCYGEAL
I CONUS MEDULLARIS
J SUBARACHNOID
SPACE
K FILIUM TERMINALE
L DURAL SAC TERMINUS

A

B

C

D
C1-C8

E
T1-T12

F
L1-L5

G
S1-S5

H

SPINAL CORD
INTERNAL FEATURES

Note in the illustration that the gray matter of the cord is arranged in the rough shape of an H. The thinner uprights of the gray matter are the **posterior horns** (columns) receiving the **posterior roots** of **spinal nerves**. The thicker uprights are the **anterior horns** (columns). The anterior horn motor neurons reside here, dispatching the anterior (motor) roots of the spinal nerves. The bar connecting the two sides of the gray columns is the **gray commissure** pierced by the midline **central canal**. In the thoracic, upper lumbar, and sacral segments of the cord, there are lateral projections (columns) or gray matter between the posterior and anterior gray columns; these incorporate the motor nuclei of autonomic nerves.

White matter surrounds gray matter. The mass of myelinated fibers between the posterior gray columns constitutes the posterior funiculi. The mass of fibers between the anterior gray columns is the anterior funiculi. The regions of white matter between adjacent pairs of posterior and anterior gray columns are lateral funiculi.

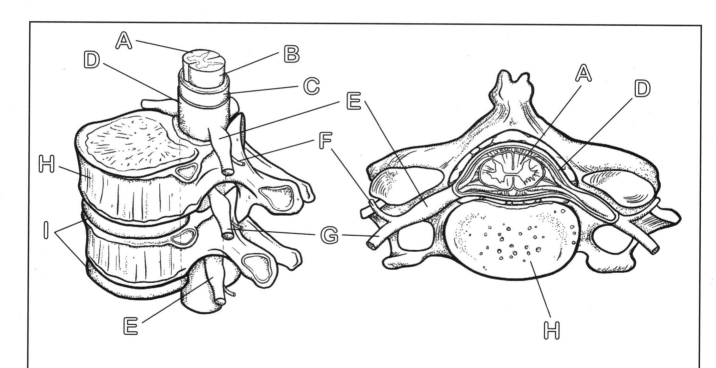

A SPINAL CORD

B PIA MATER

C ARACHNOID MATER

D DURA MATER

E SPINAL GANGLION

F POSTERIOR RAMUS

G ANTERIOR RAMUS

H VERTEBRAL BODY

I INTERVERTEBRAL DISC

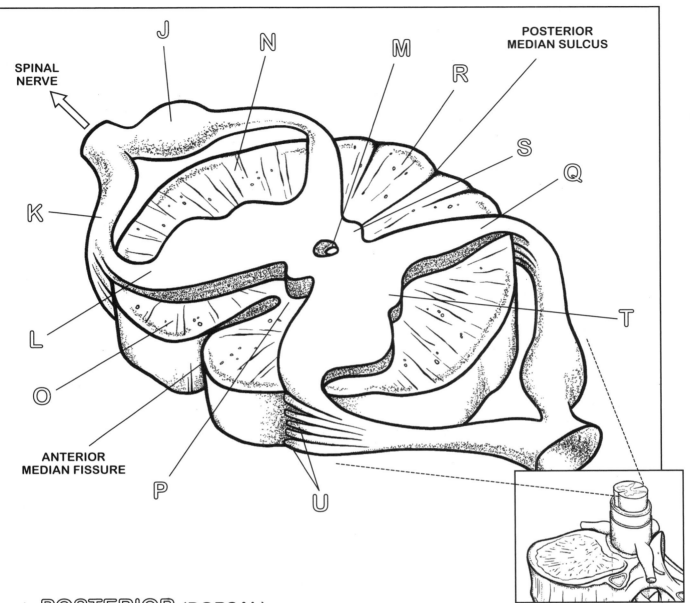

SPINAL
NERVE

POSTERIOR
MEDIAN SULCUS

ANTERIOR
MEDIAN FISSURE

J POSTERIOR **(DORSAL)**
ROOT GANGLION

K ANTERIOR **(VENTRAL)**
ROOT OF SPINAL NERVE

L ANTERIOR GRAY HORN

M CENTRAL CANAL

N LATERAL WHITE COLUMN

O ANTERIOR WHITE
COLUMN

P ANTERIOR WHITE
COMMISURE

Q POSTERIOR GRAY HORN

R POSTERIOR WHITE
COLUMN

S GRAY COMMISSURE

T LATERAL GRAY HORN

U ROOTLETS

SPINAL CORD STRUCTURAL INTEGRATION
ASCENDING PATHWAYS

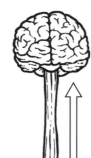

The ultimate expression of central nervous system activity is muscular contraction, which usually follows some kind of sensory input—often from receptors in skin or muscle. Thus, the common sequence is sensory input and motor output, each of which has separate pathways through the nervous system.

Sensory, or **ascending**, pathways carry impulses from the receptors into the spinal cord and up the cord through the brainstem and cerebral hemispheres to appropriate nuclei and centers for interpretation and reaction. There are other significant pathways for the cranial nerves; these, of course, bypass the spinal cord.

ASCENDING PATHWAYS

The **first-order (1°) neuron** is the term for the first link in an ascending pathway. With its receptor the 1° neuron lies outside the central nervous system—yet its cell body is in the **dorsal root ganglion** and its central process extends into the **spinal cord**. The cell body of the **second-order (2°) neuron** lies in the gray matter or, for some pathways, in the brainstem. The axon of the 2° neuron ascends to the **thalamus**. The **third-order (3°) neuron** lies in the **thalamus** and sends its axon to the **cerebral cortex**. Each link in the chain makes synaptic contact with the succeeding one.

The various sensory modalities are channeled into one of several specific tracts in the white matter of the cord. The principal ones are as follows:

• The **posterior column** is primarily responsible for transmitting position sense, movement sense, discriminative touch, and pressure. Within the posterior columns on either side of the midline are two tracts or fasciculi. The medial one is the fasciculus gracilis, the lateral one the fasciculus cuneatus.

• The **anterior spinothalamic tract** carries the sensory modalities of touch and pressure. The 1° neuron makes synaptic contact with the 2° neuron in the posterior horn of the central gray matter. The 2° neuron sends its axon across the anterior white commissure to the anterior funiculus on the other side of the midline. Here the fiber joins the anterior spinothalamic tract and ascends to the thalamus. The 3° neuron, in the thalamus, directs its axon through the internal capsule and the corona radiata to the postcentral gyrus of the cerebral cortex.

• The **lateral spinothalamic tract** (illustrated opposite) conducts pain and temperature sensations and takes the same route as the anterior spinothalamic tract, but is located in the lateral funiculus.

Other ascending tracts worthy of note but not to be examined in detail are the **spinocerebellar tracts**, conveying position and movement sense to the cerebellum; and the **spinoreticular tracts**, conducting position and movement sense to the reticular formation.

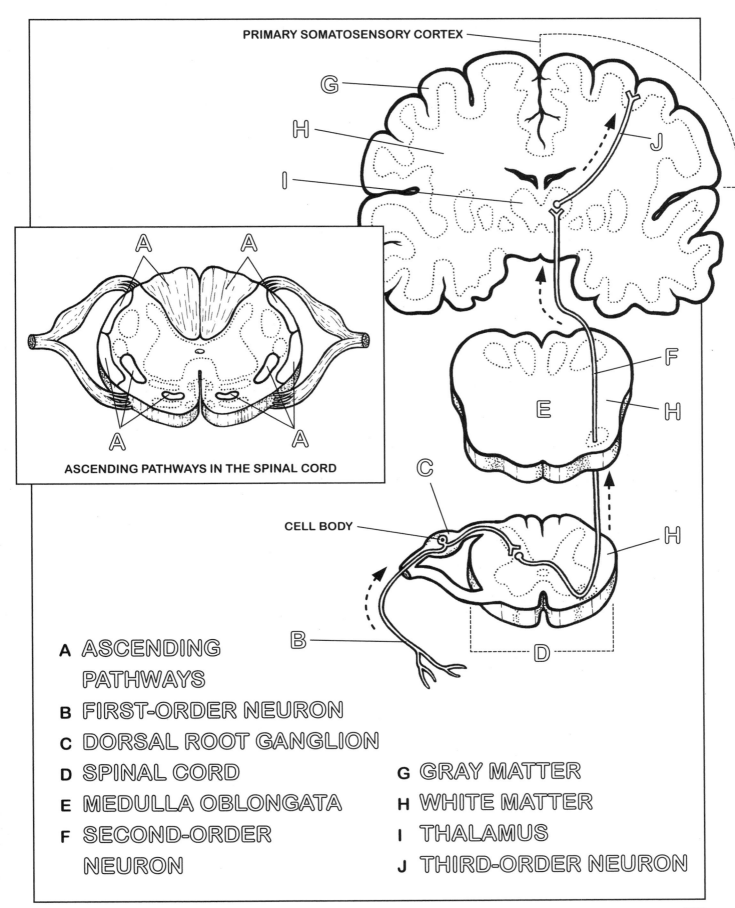

PRIMARY SOMATOSENSORY CORTEX

G

H

I

J

F

E

H

CELL BODY

C

B

H

D

ASCENDING PATHWAYS IN THE SPINAL CORD

A

A

A

A

A ASCENDING PATHWAYS

B FIRST-ORDER NEURON

C DORSAL ROOT GANGLION

D SPINAL CORD

E MEDULLA OBLONGATA

F SECOND-ORDER NEURON

G GRAY MATTER

H WHITE MATTER

I THALAMUS

J THIRD-ORDER NEURON

SPINAL CORD STRUCTURAL INTEGRATION
DESCENDING PATHWAYS

DESCENDING PATHWAYS

Motor, or **descending**, pathways are the roots for motor-related impulses from the cerebral cortex and subcortical nuclei through the brainstem (where they come under the influence of the cerebellum), down the **spinal cord** and out to the periphery to the **muscles** and glands of the body. Descending tracts may be long or short. As was said before, the ultimate expression of CNS activity is muscular contraction; you can understand that descending influences must eventually come to bear upon the motor neurons of the PNS in the brainstem or spinal cord. The axons of these neurons, the motor component of spinal nerves, represent the final common pathway to the muscles and glands. The impulses in those axons represent the algebraic summation of all cortical, brainstem, cerebellar, and spinal cord activity brought together to generate a spectrum of coordinated body movement from precise movement to rough postural changes to slight changes in the background muscular tension.

The principal descending pathways are:

• The **corticospinal** or **pyramidal tract** (illustrated opposite) is a collection of axons that carry voluntary, skilled movement and related information from the cerebral cortex to the spinal cord, with its origin primarily in the precentral gyrus of the cerebral cortex. Large multipolar neurons there give off axons which take the following course: **Cerebral cortex** (via **corona radiata**) → **internal capsule** (a white matter pathway between the thalamus and the basal ganglia) → cerebral peduncles of midbrain → pons → **anterior medulla**. Pyramidal tracts are so named because they pass through the medullary pyramids of the medulla oblongata, where corticospinal fibers converge to a point when descending from the internal capsule to the brainstem from multiple directions, giving the impression of an inverted pyramid. Some 80 to 90 percent of the axons cross there (**pyramidal decussation**) to opposite sides of the medulla and descend in the lateral funiculus as the **lateral corticospinal tract**. The uncrossed fibers pass into the anterior funiculus as the **anterior corticospinal tract**.

• **Extrapyramidal** or **nonpyramidal tracts** originate in the brainstem, carrying motor fibers to the spinal cord. These are relatively short pathways (with respect to the long corticospinal tract) from cortex to spinal cord with numerous intermediate synapses. These short fiber systems go from cortex to basal nuclei, from thalamus to basal nuclei, from basal nuclei to hypothalamus, and so on. These pathways are primarily related to the reflexes, postural adjustments, maintenance of muscle tone, and the facilitation or inhibition of voluntary movement.

To summarize, skilled movement is effected principally by the corticospinal tract; postural adjustments and other reflex movements by pathways involving the basal nuclei, thalamus, and the reticular formation; control of muscle tone and maintenance of equilibrium is regulated by cerebellar influences on midbrain and thalamus nuclei and by the vestibular nuclei of the medulla via the vestibulospinal tract. All of these influences filter down to the motor neurons that represent the final common pathways to the muscles of the body.

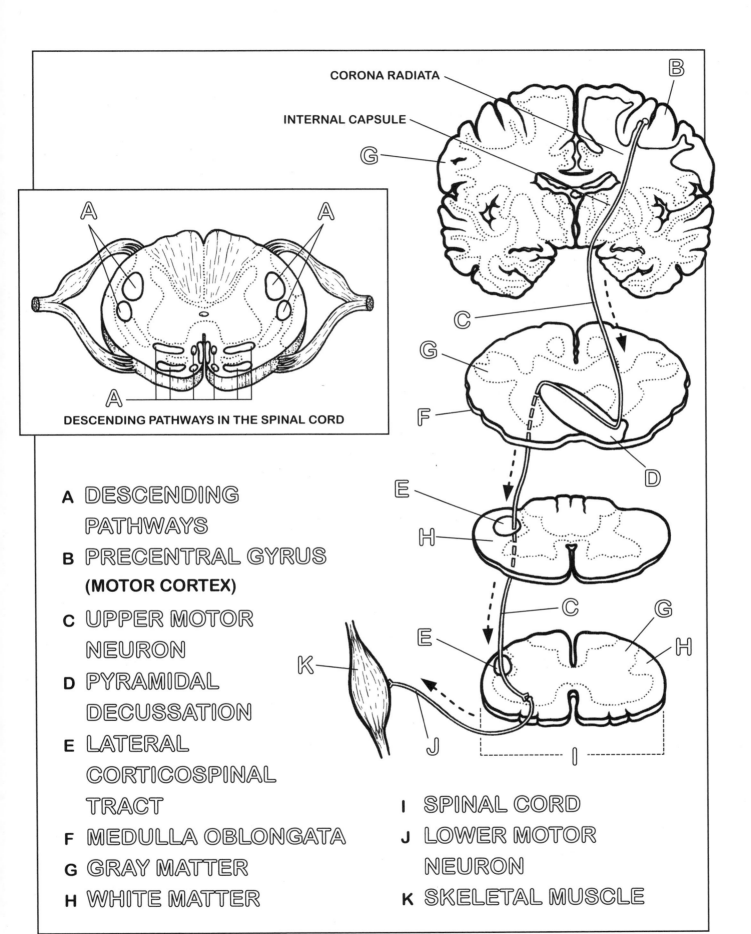

CORONA RADIATA

INTERNAL CAPSULE

DESCENDING PATHWAYS IN THE SPINAL CORD

A DESCENDING PATHWAYS

B PRECENTRAL GYRUS (MOTOR CORTEX)

C UPPER MOTOR NEURON

D PYRAMIDAL DECUSSATION

E LATERAL CORTICOSPINAL TRACT

F MEDULLA OBLONGATA

G GRAY MATTER

H WHITE MATTER

I SPINAL CORD

J LOWER MOTOR NEURON

K SKELETAL MUSCLE

ARTERIES & VEINS SERVING THE BRAIN

ARTERIES

The supply of blood to all parts of the CNS is critical. Interruptions in blood flow, and hence oxygen supply, for more than five to ten seconds can cause irreparable brain damage.

The blood supply to the brain is derived from two sources: the **vertebral arteries**, arising from the subclavian arteries; and the **internal carotid arteries**, arising from the arch of the aorta via the common carotid arteries. Note that the internal carotids terminated by giving off the middle and anterior cerebral arteries supplying the major part of the cerebral hemispheres.

An irregular ring of vessels may be seen around the optic chiasma. This is known as a **circle of Willis**, and in it the blood of the vertebrals mixes with that of the internal carotid arteries, possibly creating a "safety valve" mechanism during periods of differential blood pressure in the two pairs of arteries. The circle of Willis does not provide effective collateral circulation, however, and interruption of one of the four main arteries will seriously diminish cerebral circulation and could bring about coma and subsequent death.

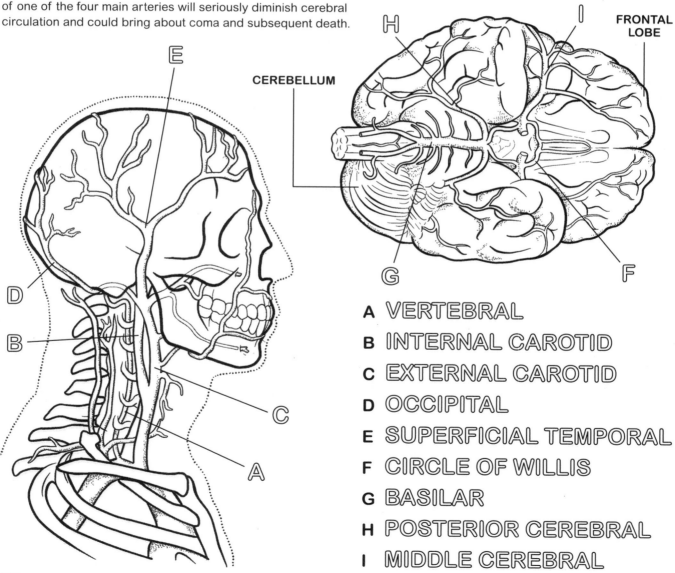

FRONTAL LOBE

CEREBELLUM

A VERTEBRAL

B INTERNAL CAROTID

C EXTERNAL CAROTID

D OCCIPITAL

E SUPERFICIAL TEMPORAL

F CIRCLE OF WILLIS

G BASILAR

H POSTERIOR CEREBRAL

I MIDDLE CEREBRAL

VEINS

The **venous drainage** of the brain is accomplished by two sets of veins, superficial and deep, with generous anastomoses between them. The veins all drain into **venous sinuses**, and some also drain into venous plexuses outside the skull.

The venous sinuses are generally bound in the periosteum of the cranial bones by the dura mater or an extension of it. They are large channels receiving venous blood from:

• Tributaries to the superficial cerebral veins;

• The internal cerebral vein, which in turn drains smaller veins of the deep cerebral hemispheres and upper brainstem.

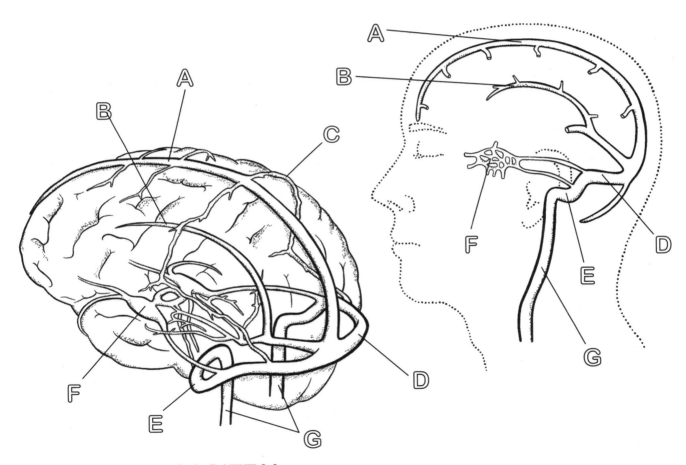

A SUPERIOR SAGITTAL SINUS

B INFERIOR SAGITTAL SINUS

C VEIN OF TROLARD

D TRANSVERSE SINUS

E SIGMOID SINUS

F CAVERNOUS SINUS

G JUGULAR

ARTERIES & VEINS
SERVING THE SPINAL CORD

Blood supply to the spinal cord is accomplished by **anterior** and **posterior spinal arteries** from the vertebral arteries. There are abundant anastomoses among these vessels and arteries along the trunk adjacent to the material column.

The veins draining the spinal cord make up complicated **plexuses** on the **anterior** and **posterior** aspect of the spinal cord. These plexuses generally drain into veins adjacent to the vertebral column, which themselves drain into the superior or inferior vena cava.

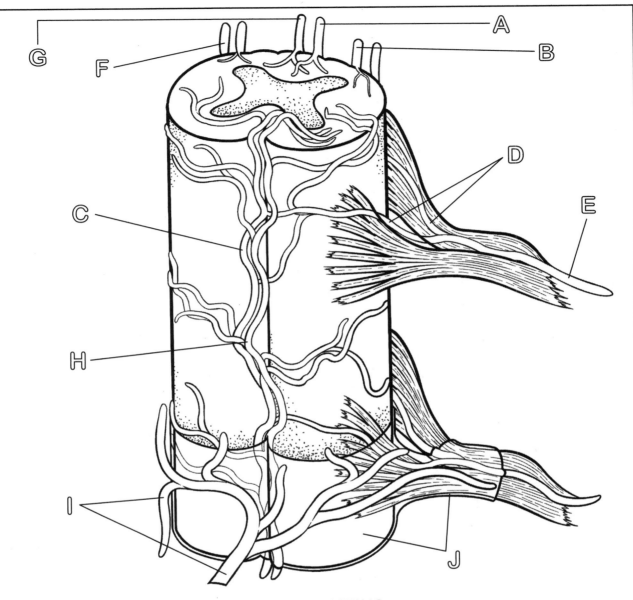

ARTERIES:

A POSTERIOR SPINAL ARTERY

B POSTEROLATERAL SPINAL ARTERY

C ANTERIOR SPINAL ARTERY

D RADICULAR ARTERY

E SPINAL ARTERY

VEINS:

F POSTEROLATERAL SPINAL VEIN

G POSTERIOR SPINAL VEIN

H ANTERIOR SPINAL VEIN

I INTERNAL VERTEBRAL VENOUS PLEXUS

J DURA MATER

MENINGES & CEREBROSPINAL FLUID

The primary function of the **meninges** and of the cerebrospinal fluid is to protect the central nervous system. The meninges consist of three connective tissue envelopes of the central nervous system. These coverings protect the brain and spinal cord in association with cerebral spinal fluid, support the brain and spinal cord within their bony housings, and serve as a vehicle for vessels supplying and draining the brain and spinal cord.

The innermost covering is closely applied to the surface of the brain and spinal cord. It is a delicate, soft layer, and has its name: **pia mater**. It cannot be easily separated from the underlying nervous tissue, as it is believed to send projections into it.

The pia embraces the CNS in a continuous sheet except where three small foramina penetrate it in the roof of the fourth ventricle and where the many blood vessels supplying the brain and spinal cord pass through. The pia terminates caudal to the conus terminale as the filum terminale.

The middle layer of the meninges is a filmy, web-like membrane, termed the **arachnoid**. It is separated from the underlying pia by a space that is crossed by small beams of connective tissue (**trabeculae**). This **arachnoid space** conducts the cerebrospinal fluid. The arachnoid ends caudal to the conus terminale as the internal lining of the dural sac.

The **dura mater** is the tough, fibrous outer covering of the CNS. It consists of two layers: the outer **periosteal layer**, serving as the periosteum of the cranial bones and terminating at the foramen magnum; and the inner **meningeal layer**, separated from the arachnoid by a subdural space. Lying within the two layers are the **dural sinuses**. The inner layer of the dura dips into the great crevices between certain parts of the brain, forming dividers or septa.

This dura is a vascularized structure, being supplied by the middle meningeal artery and its branches, which leave an impression on the inner surface of the calvaria. This important artery, a second-order branch of the external carotid artery, is subject to hemorrhage with fractures of the skull in the region—hemorrhages that are frequently fatal (epidural hemorrhage). The dura terminates caudally as a sac at the S2 vertebral level.

The spinal cord, continuous with the brain, is also surrounded by the three meninges. The dura mater extends from the foramen magnum to the sacrum and coccyx.

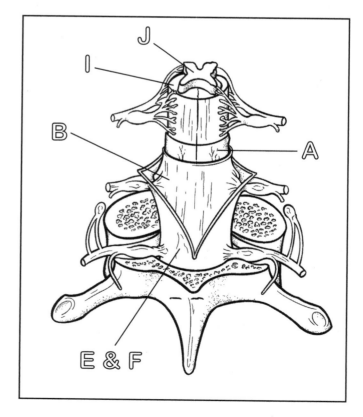

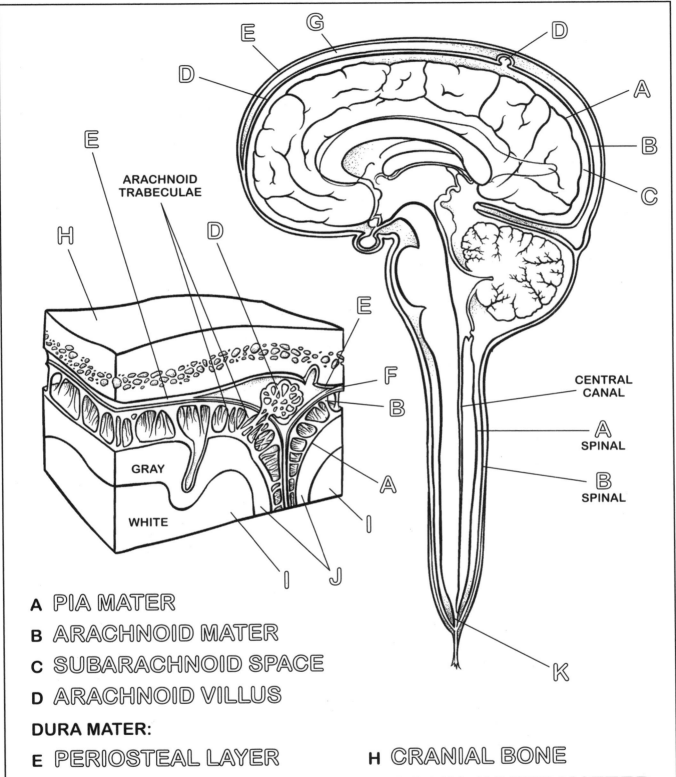

ARACHNOID
TRABECULAE

GRAY

WHITE

CENTRAL
CANAL

A SPINAL

B SPINAL

A PIA MATER

B ARACHNOID MATER

C SUBARACHNOID SPACE

D ARACHNOID VILLUS

DURA MATER:

E PERIOSTEAL LAYER

F MENINGEAL LAYER

G SUPERIOR SAGITTAL
SINUS

H CRANIAL BONE

I BRAIN: WHITE MATTER

J BRAIN: GRAY MATTER

K FILIUM TERMINALE

SECTION SEVEN:
SENSORY RECEPTORS & ASSOCIATED ORGANS OF THE HEAD

Throughout evolutionary history, animals have consistently greeted their environment head first. This is not by accident, for the head is gifted with a set of exteroceptors unlike any other in the body. Consider: A wildcat apprehensively checks out a cave as potential quarters for his mate. Approaching the entrance, his nose wrinkles, nostrils flared for suspicious scents; his ears prick up smartly and turn toward the cave, surveying every sound; his dilated eyes dart from point to point, seeking movement of the slightest sort.

Preservation at stake, do we not emulate that very pattern? When investigating new ground, do we not crouch forward—head first—using our receptors of sight, sound, and smell to inform ourselves of the state of things?

These special sensory receptors are:
• Visual sensors located in the globe of the eye.
• Auditory and head position sensors in the inner ear cavity of the temporal bone.
• Olfactory senses in the roof of the nasal cavity.
• Gustatory sensors in the surface lining of the tongue.

VISUAL RECEPTORS
THE EYE

If one were to design a functioning visual receptors system, one would have to include in the plans a number of structures. The full list would be:

- **Protective housing**
- **Photoreceptors**
- **Lightproof shroud**
- **Refractive media**
- **Light-regulating and distance accommodation mechanism**
- **Source of power: nutrition and innervation**
- **Mechanism for movement**
- **Mechanism for maintaining the external surface**

All can be found in the eye, a fascinating and complex component of the human body, and will be explored fully on the following pages.

The eye and related structures are firmly entrenched within the deep recesses of the bony skull known as the **orbit**, protected circumferentially by packages of **fat** and held firmly in place by extrinsic **muscular** and **ligamentous** structures. The front visible part of the eye has as its outer layer a dense, fibrous, rubber-like housing, the **sclera**. It is whitish and quite resilient.

As you may assume, eyes are part of the sensory nervous system. To detect visible light, they have **rod** and **cone photoreceptor cells**. Rod cells are responsible for vision in low-light conditions, while cone cells are responsible for bright light and color vision. A human eye contains about 120 million rod cells and 6 million cone cells!

When stimulated by light these cells generate and transmit impulses via the **optic nerve** to the **optic chiasma**—a point of both union and decussation, where right and left visual information crosses to opposite sides of the brain. At the end of each **optic tract**, retinal nerves synapse with other visual pathway nerves in the **lateral geniculate body (LGB)** of the **thalamus**; and finally the **visual cortex** of the **occipital lobe** recieves the information. This in turn facilitates the perception of color, shape, depth, and movement—it is at the occipital lobe where the visual impulses are interpreted and an image is created.

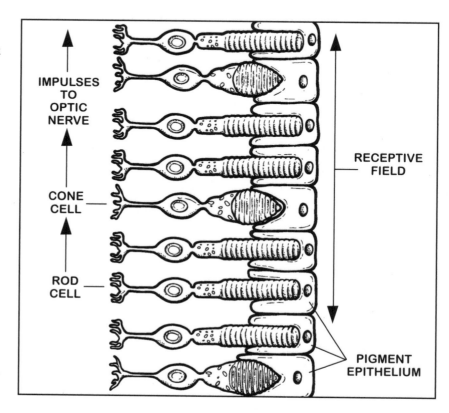

IMPULSES TO OPTIC NERVE

CONE CELL

ROD CELL

RECEPTIVE FIELD

PIGMENT EPITHELIUM

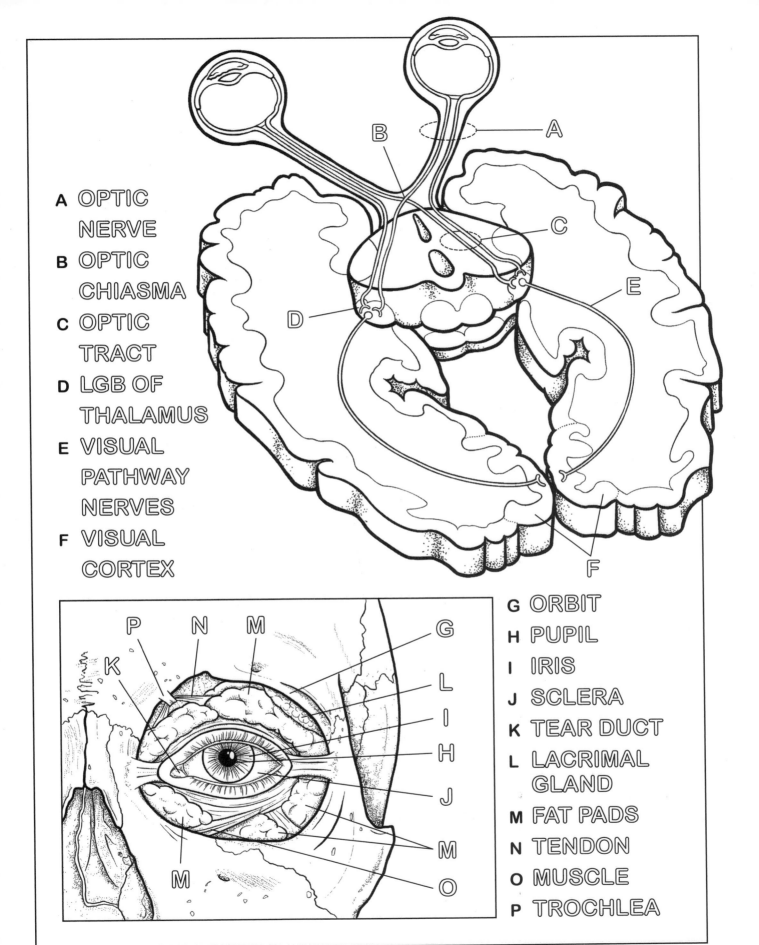

A OPTIC NERVE
B OPTIC CHIASMA
C OPTIC TRACT
D LGB OF THALAMUS
E VISUAL PATHWAY NERVES
F VISUAL CORTEX

G ORBIT
H PUPIL
I IRIS
J SCLERA
K TEAR DUCT
L LACRIMAL GLAND
M FAT PADS
N TENDON
O MUSCLE
P TROCHLEA

VISUAL RECEPTORS AND RELATED STRUCTURES
ANATOMY OF THE EYE

LIGHT-REGULATING AND DISTANCE ACCOMMODATION STRUCTURES

Light entering the eyeball must be carefully directed. Thus a dark shroud coats the inner surface of the **sclera**—the highly pigmented and extremely vascular **choroid** absorbs light and helps to limit any reflections in the eye that could otherwise impair vision. The anterior extension of the choroid is the **ciliary body**, which resembles the pedals of a daisy when viewed from behind. The inner border of the ciliary body is thrown into a series of folds—the **ciliary processes**. Within the ciliary body are **ciliary muscles**, smooth muscle fibers that alter the shape of the **lens**.

The anterior and medial radial projections from the ciliary body are part of the disc-like **iris**, whose posterior surface is lined with a richly pigmented layer. The iris has two sets of muscle fibers: the **radial dilator pupillae muscles** and the more circular **sphincter pupillae muscles**. These muscles dilate or constrict the diameter of the **pupil**, the central aperture of the iris, which regulates the amount of light from the outside world to the **retina**, a complex ten-layered structure and the inner layer of the posterior two thirds of the eyeball. One of the deeper layers is occupied by the aforementioned photoreceptor cells. Generally the organization of the retina is uniform throughout, with two exceptions: an indented, thinned area densely populated with cones and entirely without rods known as the **fovea centralis**. This is the region of sharpest visual acuity. The other nonuniform region of the retina is the **optic disc**, where optic tract fibers converge to form a bundle and leave the globe of the eye posteriorly. There are no photoreceptors here, so a "blind spot" is created.

REFRACTIVE MEDIA

To concentrate light waves at a focal point on the retina, they must be bent or **refracted** as they pass through the eye. This is accomplished by a number of structures, described in order of contact with incoming light:

The **cornea** is the anterior, transparent continuation of the sclera. Composed of a lamellae of connective tissue bounded by epithelium, the cornea is extremely sensitive—touching it will set off the familiar blink reflex. The cornea is the principal refractive medium of the eye. Immediately deep to the cornea is the **aqueous humor**, a chamber filled with a watery (aqueous) fluid (humor). This anterior chamber is in communication with the **posterior chamber** via a small channel between the iris and lens. Aside from acting as a refractive medium, the aqueous humor places uniform pressure on the eyeball interior, helping to maintain its roundness.

Just deep to the anterior chamber and suspended by **suspensory ligaments** arising largely from the ciliary body is the **lens**. This encapsulated epithelial body is a flexible structure (unlike the lenses in glasses). Functioning to refract light waves to a focal point on the retina, the lens can alter its shape in accommodation of near or far vision. Immediately deep to the lens and filling the posterior two thirds of the eyeball interior is the gelatinous **vitreous body**. Its interior border supports the lens and contributes fibers toward suspensory ligaments of the lens.

SOURCE OF NUTRITION

The primary source of blood to the orbital contents is the ophthalmic artery, a branch of the internal carotid artery. It enters the orbit through the optic canal and branches extensively. The artery to the retina is the central artery, traveling in the substance of the **optic nerve**. Generally, the eyeball structures are serviced by the ciliary arteries, all of them branches to the ophthalmic artery.

This drainage of the iris, retina, and related structures is accomplished by tributaries of the vorticose veins. These latter veins drain into the cavernous sinus via ophthalmic and ciliary veins.

The critical blood supply of the retina is derived from vessels of the choroid.

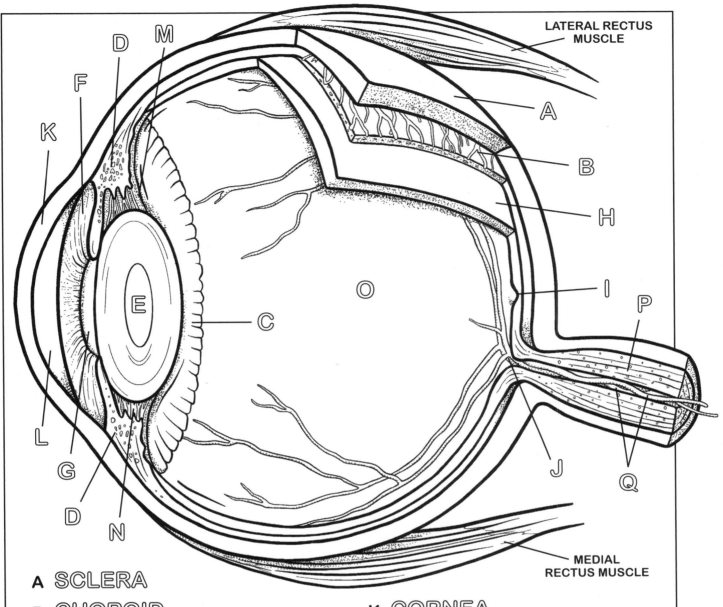

LATERAL RECTUS MUSCLE

MEDIAL RECTUS MUSCLE

A SCLERA

B CHOROID

C CILIARY BODY

D CILIARY MUSCLE

E LENS

F IRIS

G PUPIL

H RETINA

I FOVEA CENTRALIS

J OPTIC DISC

K CORNEA

L AQUEOUS HUMOR

M POSTERIOR CHAMBER

N SUSPENSORY LIGAMENTS

O VITREOUS BODY

P OPTIC NERVE

Q RETINAL BLOOD VESSELS

VISUAL RECEPTORS AND RELATED STRUCTURES
THE EYE: MYOLOGY & INNERVATION

MECHANISMS FOR EYE MOVEMENT

Movement of the orbit is a function of six slender skeletal muscles inserting on the sclera which (with two exceptions) arise from a common circular tendon ringing the optic canal. The function of the extrinsic muscle is often complex—no one muscle acts alone.

Note that four of the muscles are straight and are so named: **rectus**. Two are angled with reference to the straight muscles and are also named: **oblique**. The names of the muscles are prefixed with the term of orientation, e.g., **superior**, **inferior**. The **trochlea** here is a pulley-like structure for the tendon of the **superior oblique** muscle, which passes though it.

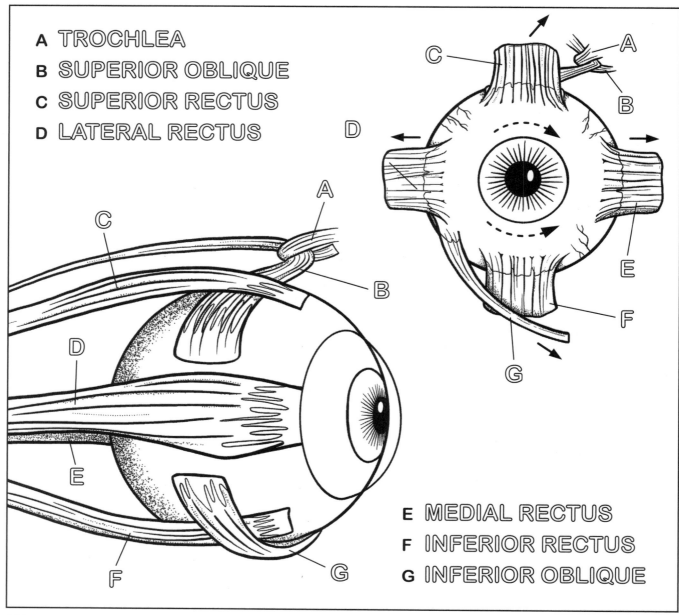

A TROCHLEA
B SUPERIOR OBLIQUE
C SUPERIOR RECTUS
D LATERAL RECTUS

E MEDIAL RECTUS
F INFERIOR RECTUS
G INFERIOR OBLIQUE

INNERVATION OF ORBITAL STRUCTURES

The nerves relating to the orbit and its occupants can be divided into three functional categories:

Sensory: the optic nerve conducting visual impulses, and the other nerves (third, fourth, fifth, and sixth cranial nerves) conducting impulses of pain, pressure, in proprioceptive stimuli.

Autonomic: the parasympathetic preganglionic nerves riding with the oculomotor (third) nerve to the **ciliary ganglion**; parasympathetic postganglionic nerves to the ciliary (accommodating) muscles and pupillary sphincter muscle; sympathetic postganglionic nerves to the blood vessels and dilator muscle of the pupil.

Somatic motor: the oculomotor nerve facilitates movement of all eye muscles except the lateral rectus, which is innervated by the abducens (sixth) nerve, and the superior oblique, which is supplied by the trochlear (fourth) nerve.

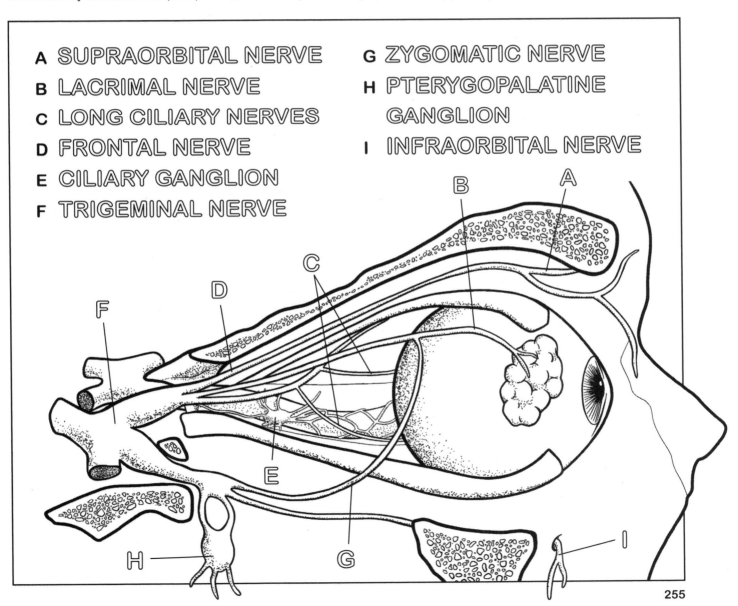

A SUPRAORBITAL NERVE

B LACRIMAL NERVE

C LONG CILIARY NERVES

D FRONTAL NERVE

E CILIARY GANGLION

F TRIGEMINAL NERVE

G ZYGOMATIC NERVE

H PTERYGOPALATINE GANGLION

I INFRAORBITAL NERVE

VISUAL RECEPTORS AND RELATED STRUCTURES
THE EYE: EXTERNAL SURFACE

MAINTENANCE OF EXTERNAL SURFACE OF EYE

The cornea and anterior sclera are in constant contact with airborne pollutants of the external environment. Demands are put upon the protective elements of these surfaces:

- The **eyelids**
- The **conjunctiva**
- The **lacrimal** apparatus

The eyeballs are protected by two mobile **eyelids** (palpebrae). Eyelids are fibrous connective tissue structures incorporating a very thin layer of skin externally, a mucous membrane (conjunctiva) on its internal surface and groups of smooth and skeletal muscle and sebaceous-like (tarsal) glands within. The eyelids protect the eye against such missiles as splintering wood, metal shavings, or just plain dust. With each closing of the lids, the cornea gets a "washing" much as a car window is washed by a window wiper blade.

The **conjunctiva** is the thin, vascular mucous membrane lining the internal surfaces of the eyelids and **reflecting** onto the cornea and anterior sclera. The conjunctiva plays an important role in protecting the cornea from physical contact with damaging particles. Perhaps most importantly, it keeps the cornea moist— a prerequisite to corneal transparency.

The upper reflection of the conjunctiva contains several ducts of the secretory **lacrimal gland**. Occupying the anterolateral corner of the orbit just above and deep to the upper eyelid, the lacrimal gland reposes on the lateral and superior rectus muscles. Some dozen ducts drain this autonomically innervated tear gland.

Lacrimal glands perform the obvious function of moistening and cleansing. In certain emotional situations, the lacrimal gland is stimulated to secrete—and the excess tears overflow the lids.

A LACRIMAL GLAND

B DUCTS OF
  LACRIMAL GLAND

C UPPER EYELID

D LOWER EYELID

E CONJUNCTIVA

F REFLECTION OF
  CONJUNCTIVA

G EYE

H LACRIMAL SAC

I NASOLACRIMAL
  DUCT

J AMPULLA OF
  LACRIMAL DUCT

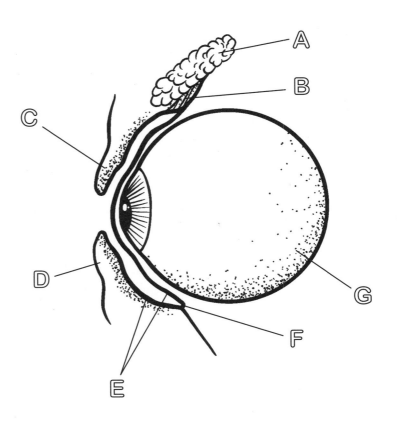

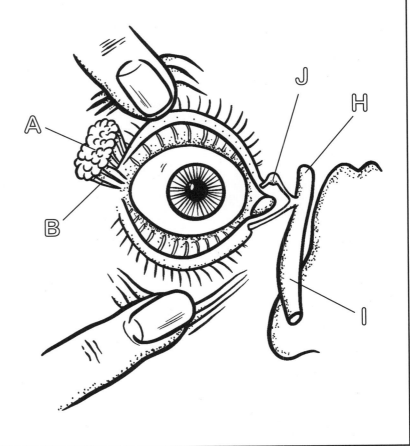

AUDITORY RECEPTORS
THE EAR

There are no sensory receptors in the nervous system specifically for sound energy, and therein lies the basis for the complexity of the **ear**. Sound waves collected by these elaborate appendages are funneled into the cavernous interior of the temporal bone and converted into mechanical energy. Then, still deeper, this energy is translated into nervous impulses, a currency that the CNS can handle. Also in the internal ear are the vestibular cells for sensing head movements.

That flap of skin protruding from each side of your head is more precisely termed the **auricle**. The "ear" is the entire audiovestibular apparatus. Anatomically it is divided into three areas:

External ear: The orifice at the auricle projects medially via a funnel-like apparatus that ducts air back to the tympanic membrane. This auditory canal or **external auditory meatus** is supported by the **temporal bone**, and it terminates at the entrance to the . . .

Middle ear: Sealed by the **tympanic membrane** or "eardrum," this cavity houses bony and muscular structures that convert sound energy into mechanical energy and transmit it to the chambers of the . . .

Internal ear: The home of sensory organs of audition and equilibrium. From here the vestibulocochlear nerve (eighth cranial nerve) conducts auditory and vestibular-related impulses to the lower brainstem for interpretation.

The middle ear, or **tympanic cavity,** is a region that may be likened to a room. It has a roof, a floor, four walls, two windows, and a carpet of mucous membrane. It even has a front and back "door," and it is furnished with bony ligamentous, muscular, and neural "furniture." The roof (**tegmen tympani**) constitutes part of the petrous portion of the temporal bone. The floor (**jugular wall**) consists of the lower plate of the petrous portion of the temporal bone. It overlies the internal jugular vein. The lateral wall is the **tympanic membrane**. The medial wall of the tympanic cavity is a lateral wall of the internal ear; penetrating it are an **oval window**, with a membranous "windowpane," and a **round window**, also sealed by a membrane. The posterior wall contains the epitympanic recess (the back "door"), which communicates posteriorly and inferiorly with the mastoid air cells. The anterior wall (**carotid wall**) carries the impression of the internal carotid artery, which passes alongside the middle ear as it ascends to the brain. In the anteromedial corner of this wall is an orifice, the termination of the pharyngotympanic (auditory) tube (the front "door").

Three small articulated bones known as the **auditory ossicles** bridge the middle ear cavity between the tympanic membrane and the oval window. They articulate with each other by means of freely movable synovial joints. The **malleus** attaches by its handle to the tympanic membrane. Its neighbor is the **incus**, which attaches to the head of the **stapes**. This stapes is joined to the oval window by its firmly embedded footplate. Thus vibrations received by the stapes from its fellow ossicles produce piston-like movements in the oval window and so set the fluid inside in motion. Slender skeletal muscles are attached to two of the ossicles from origins on the wall of the middle ear: The **tensor tympani** attaches to the handle of the malleus, and the stapedius inserts on the head of the stapes.

The **eustachian** or **pharyngotympanic tube** extends from the middle-ear cavity to the nasopharynx. It is carpeted with mucous membrane and serves to equalize the air pressure between the two cavities. It is sometimes called the auditory tube.

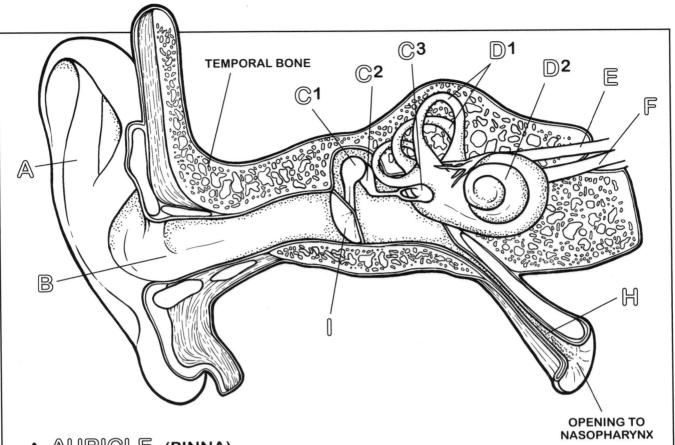

TEMPORAL BONE

A AURICLE (PINNA)

B EXTERNAL AUDITORY MEATUS

C MIDDLE EAR (TYMPANIC CAVITY)

C1 MALLEUS (HAMMER)

C2 INCUS (ANVIL)

C3 STAPES (STIRRUP)

D INNER EAR

D1 SEMICIRCULAR CANALS

D2 COCHLEA

E VESTIBULAR NERVE

F COCHLEAR NERVE

G OVAL & ROUND WINDOWS

H EUSTACHIAN TUBE

I TYMPANIC MEMBRANE

J TENSOR TYMPANI

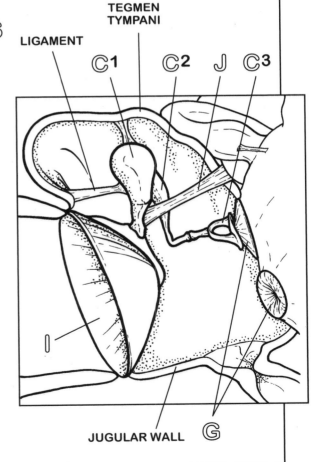

OPENING TO NASOPHARYNX

TEGMEN TYMPANI

LIGAMENT

JUGULAR WALL

AUDITORY SYSTEM & RELATED STRUCTURES
INTERNAL EAR

The internal ear is a series of fluid-filled bony chambers (**bony labyrinth**) housing a series of fluid-filled membranous chambers (**membranous labyrinth**).

THE BONY LABYRINTH

A series of passageways in the temporal bone, the bony labyrinth is a closed set of channels and contains **perilymph**, an extracellular fluid. The bony labyrinth consists of:

• A **vestibule** opening into both the **semicircular canals** and the scala vestibuli of the cochlea. It is also in communication with the **oval window.**
• Three **semicircular canals.**
• The **cochlea**, consisting of the **scala vestibuli**, **scala tympani**, and housing the membranous **cochlear duct.**

THE MEMBRANOUS LABYRINTH

This is an epithelium-lined series of structures within the bony labyrinth, although it is not in communication with the bony labyrinth. It includes the large **utricle** and the smaller **saccule**, both within the **vestibule**; the **semicircular ducts** within the bony **semicircular canals**; and the **cochlear duct**, more or less centered within the cochlea. The membranous labyrinth is filled with **endolymph**, a fluid similar to intracellular fluid—as is the **endolymphatic sac.**

THE COCHLEA

Deep to the **oval window** is the **vestibule**, a chamber that opens to the **scala vestibuli**, a canal within the bony cochlea. Thus the oval window is in direct communication with the scala vestibuli. The scala vestibuli spirals two and one half times around the bony core, or **modiolus**, of the cochlea. Just below it is another canal in the bone of the cochlea, the **scala tympani**. The two canals communicate only at the top of the spiral, through a hole, the **helicotrema**. Thus the one connection between the oval window and the scala tympani is through the scala vestibuli.

Lying between the two canals is the smaller **cochlear duct**. It is part of the membranous labyrinth, hence surrounded by membrane. Triangular in cross section, the cochlear duct may be likened to a long tent, placed at the modiolar edge of the scala vestibuli and following its course all the way to the helicotrema. The cochlear duct is closed at both ends, has a delicate membranous (vestibular) roof, has a vascular and ligamentous lateral wall, and has a partly bony, partly membranous floor projecting from the core of the cochlea. A complicated and specialized structure arising from the floor of the duct is the **organ of Corti**, composed of sensitive **hair cells** and **supporting cells**, and an underlying tectorial membrane arising indirectly from the osseous (spiral) lamina. The organ of Corti is the receptor organ for hearing. It converts sounds into electrical signals that are then transmitted to the brainstem via the **cochlear** and **auditory nerves**.

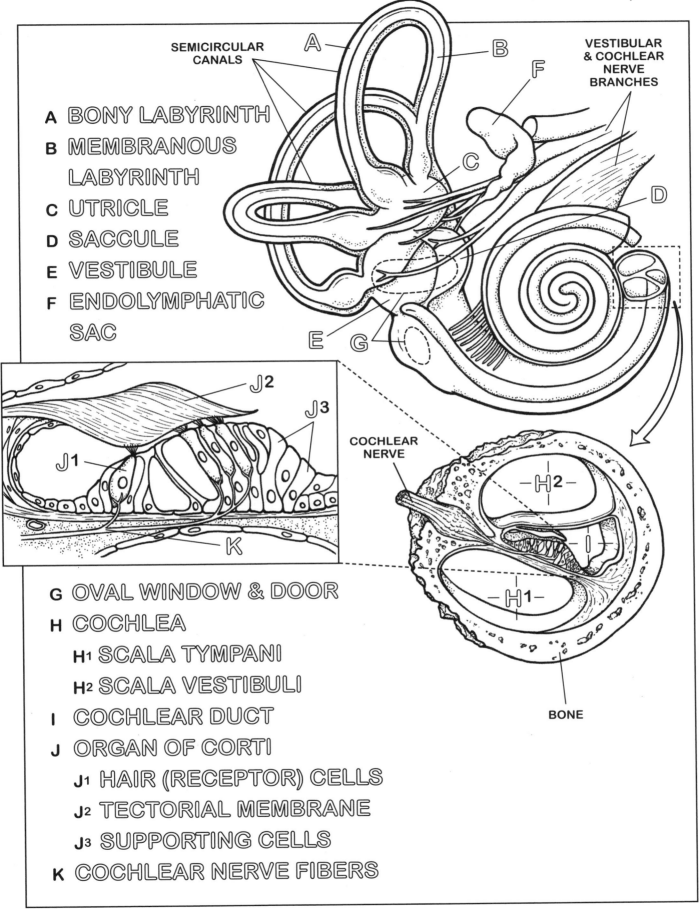

A **BONY LABYRINTH**

B **MEMBRANOUS LABYRINTH**

C **UTRICLE**

D **SACCULE**

E **VESTIBULE**

F **ENDOLYMPHATIC SAC**

SEMICIRCULAR CANALS

VESTIBULAR & COCHLEAR NERVE BRANCHES

COCHLEAR NERVE

BONE

G **OVAL WINDOW & DOOR**

H **COCHLEA**

   H¹ **SCALA TYMPANI**

   H² **SCALA VESTIBULI**

I **COCHLEAR DUCT**

J **ORGAN OF CORTI**

   J¹ **HAIR (RECEPTOR) CELLS**

   J² **TECTORIAL MEMBRANE**

   J³ **SUPPORTING CELLS**

K **COCHLEAR NERVE FIBERS**

AUDITORY SYSTEM & RELATED STRUCTURES
VESTIBULAR STRUCTURES

EQUILIBRIUM
The structures of the internal ear associated with equilibrium are:

• The three bony semicircular canals housing the **semicircular ducts**.
•The larger utricle and the smaller saccule—as shown on the previous page, both are membranous structures within the bony vestibule.
• The pressure-relief valve of the internal ear, the endolymphatic duct, which may be involved in fluid reabsorption.

Hair cells embedded in **endolymph**, a gelatinous material, lie in the **ampullae** (dilated ends) of the semicircular ducts. The hair cells are sensory receptors of the vestibular portion of the eighth nerve. When movement of the head causes the endolymph in the appropriate duct to move, the **stereocilia** topping the hair cells bend, stimulating neurons which then alert the brain that the head is tilted. In the utricle and saccule too, hair cells are embedded in a gelatinous material. Moving the head also forces these top-heavy sensory receptor cells to bend and fire an impulse via the vestibular portion of the eighth nerve. Thus any movements of the head stimulate one or more of these sensory cell patches to fire afferent impulses.

The architecture of the **anterior**, **lateral**, and **posterior canals**—oriented in three different planes—is such that all movements of the head can be detected. The impulses are distributed to the brainstem, cerebellum, and spinal cord.

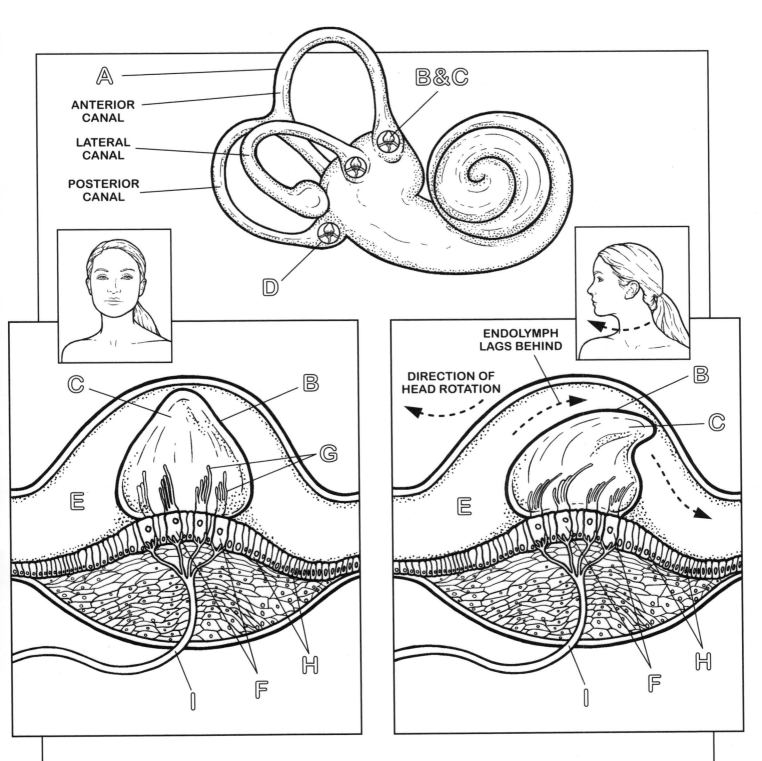

A ANTERIOR CANAL

LATERAL CANAL

POSTERIOR CANAL

B&C

D

ENDOLYMPH LAGS BEHIND

DIRECTION OF HEAD ROTATION

A SEMICIRCULAR DUCTS
B CRISTA AMPULLARIS
C CUPULA
D AMPULLAE
E ENDOLYMPH
F HAIR CELLS
G STEREOCILIA
H SUPPORTING CELLS
I SENSORY NERVE FIBERS

GUSTATORY RECEPTORS
THE TONGUE

Since the things we require for maintenance of our bodies normally enter by way of the mouth, it is no surprise that there is a taste receptor just inside. Some nine thousand chemoreceptors are located on the tongue and pharynx. The receptors look like the buds of a flowering plant and, therefore, are captioned **taste buds**, or **gustatory** (taste) **organs**.

Taste buds are epithelial structures (less than one mm in height) located on the walls of the pharynx and palate to a lesser degree and on the **circumvallate papillae** of the tongue to a greater degree.

Each taste bud is comprised of fifty to a hundred and fifty **taste receptor cells**, which discriminate among four chemical modalities: sweet, sour, bitter, and salty. Some receptors are sensitive to more than one mode, but probably never all four. Substances placed in the mouth dissolve sufficiently with the help of saliva to bathe the crypts of the **papillae** and the taste receptor cells, which have protrusions called **gustatory hairs** that are stimulated by food molecules. The hairs and cells then send an electrical impulse to the brain via **afferent taste nerves** for interpretation. Taste receptor cells are short-lived and replaced about every ten days, since they are easily damaged by the activities that occur in the mouth. The **basal cells** divide to form new taste receptor cells.

There is no specific cortical area for taste reception. Interpretation is made at the sensory cortex subserving the facial region.

The phenomenon of "discriminating taste" is based on a different mixture of (1) the four basic tastes, (2) the temperature and texture of the food, and (3) the smell of the food.

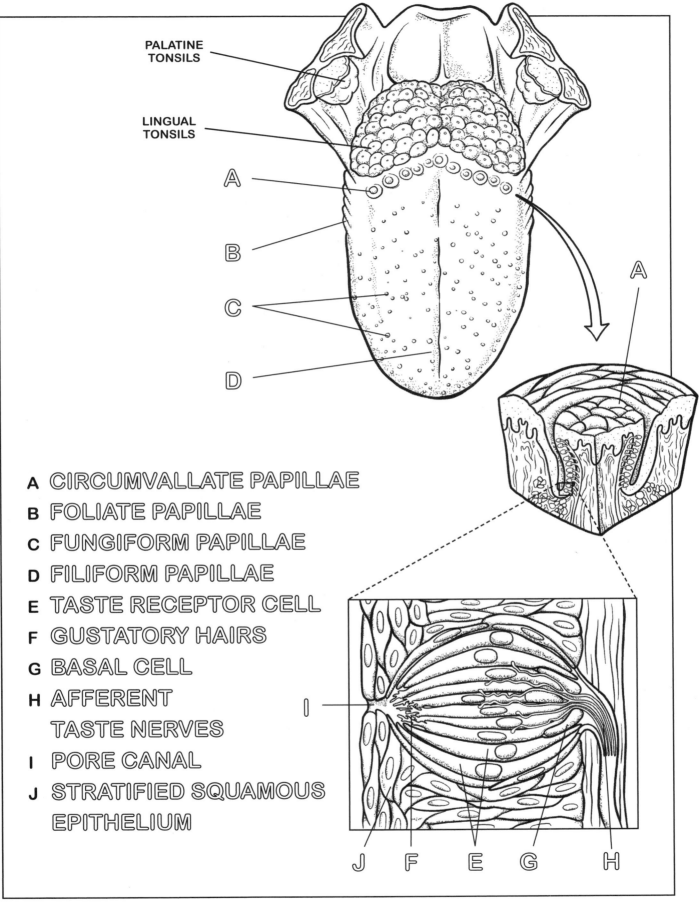

PALATINE TONSILS

LINGUAL TONSILS

A

B

C

D

A

A CIRCUMVALLATE PAPILLAE
B FOLIATE PAPILLAE
C FUNGIFORM PAPILLAE
D FILIFORM PAPILLAE
E TASTE RECEPTOR CELL
F GUSTATORY HAIRS
G BASAL CELL
H AFFERENT TASTE NERVES
I PORE CANAL
J STRATIFIED SQUAMOUS EPITHELIUM

I

J   F   E   G   H

OLFACTORY RECEPTORS
THE NOSE

In evolutionary terms, the sense of smell, or **olfaction**, has great significance—particularly as it relates to the development of instincts and emotions. In less complex animals, the part of the brain (rhinencephalon) devoted to interpretation of smell is quite large. In humans, no part of the rhinencephalon is devoted to olfaction, as all of it is taken up by neuronal circuits to emotional behavior.

Wedged between **supporting cells** of olfactory epithelium, **olfactory receptor cells** are located in the nasal mucosa at the roof of the nasal cavity. These neurons have **cilia** that project into the mucous covering the **nasal mucous membrane** and contain olfactory receptors sensitive to certain chemicals dissolved in the mucous. Odor molecules will bind to the olfactory receptors, initiating an electrical signal that travels to the **olfactory bulb** lying atop the **cribriform plate** of the **ethmoid bone** in the anterior cranial fossae. From there, the impulses pass via the olfactory stria to the underside of the temporal and frontal lobes for processing.

The basis for olfactory discrimination is unknown. When odors pass into the nose, they generally pass straight through into the nasopharynx. When one wants to partake of an odor in greater magnitude, one dilates the nares, tilts the head back to expose the sensory cells directly, and inhales. This activity is more commonly called sniffing!

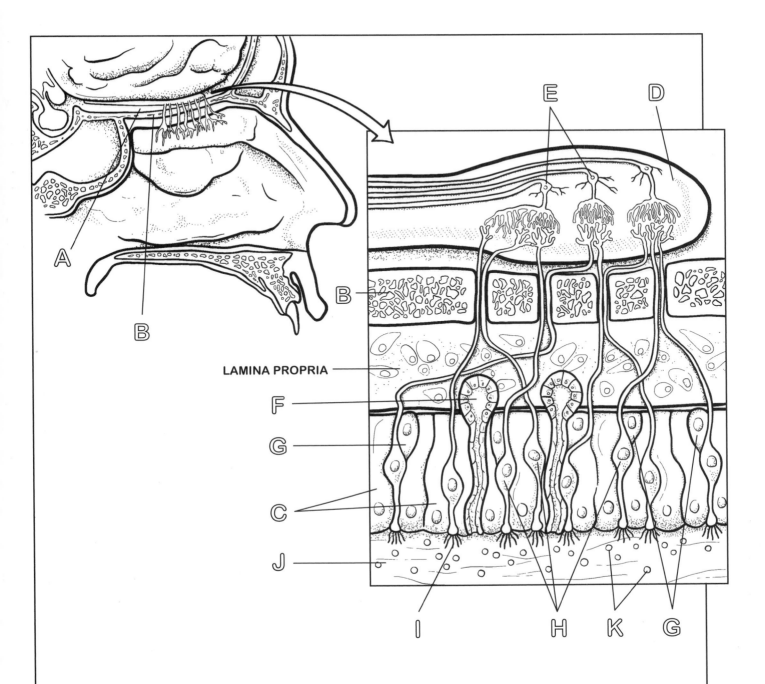

LAMINA PROPRIA

A OLFACTORY TRACT

B CRIBRIFORM PLATE
OF ETHMOID BONE

C SUPPORTING CELLS

D OLFACTORY BULB

E PROJECTION NEURON

F OLFACTORY GLAND

G BASAL CELL

H OLFACTORY
RECEPTOR CELL

I CILIA

J NASAL MUCOUS
MEMBRANE

K ODOR MOLECULES

SECTION EIGHT:
THE BODY WALL

You have just completed a study of the most complicated regions of the body: the head and neck.
We now move on to studying the body wall.

The body wall consists of the skin, fascia, and musculoskeletal structures serving as boundaries to the trunk.
It includes the vertebral column, the ribs and sternum, intercostal musculature, the diaphragm,
muscles of the abdominal wall, and muscles of the back.

Interrelations among these structures permit us to demonstrate the following regions or cavities:

Thorax: bounded by the sternum and costal cartilages anteriorly, the ribs antero- and posterolaterally, the vertebral
column posteriorly, and the diaphragm below. The intercostal (between-the-ribs) musculature and muscles of
the deep back contribute to these bony boundaries. The space created within constitutes the thoracic cavity and
incorporates the lungs, heart, and related structures.

Abdomen: bounded by the abdominal musculature anteriorly and laterally, by the thoracic diaphragm and lower ribs
above, by the posterior abdominal and deep back musculature posteriorly, and continuous with the pelvis below.
The abdominal cavity created within contains the alimentary canal and related viscera, the spleen, the kidneys, and
assorted endocrine glands, ducts, vessels, and nerves—all topics of interest and explained in more detail in other
sections of the book.

The anterior abdominal wall contributes to the boundaries of the pelvis, a cavity continuous above with the
abdominal cavity, surrounded largely by bone and containing certain structures of the digestive, urinary,
and reproductive systems.

VERTEBRA ANATOMY

A column of bones in your back constitutes the basic pillar that our body depends upon for support. A **vertebra** is one of the bony segments that form this pillar—vertebra are the building blocks of the spine.

A typical vertebra basically consists of a **body** and **vertebral arch**, the latter so called because it arches over the spinal cord. The vertebral arch has two sides—the **pedicles** and the **laminae**. Projecting from the junction of the pedicle and the lamina on each side is a process, the **transverse process**. These are, quite simply, "muscle hangers." As many as sixteen different muscles may insert or arise from one lumbar transverse process! In the thoracic region each transverse process gives attachment to a rib. Also projecting from the junction of the pedicle and the lamina on each side but directed inferiorly and superiorly are **articular processes**, which articulate with those of the adjacent vertebra. These are synovial joints. A **spinous process** or spine projects posteriorly from the fused lamina.

When the vertebrae are in articulation, two holes become visible:
• a passageway for the spinal cord created by the vertebral arches in series: the **vertebral canal**;
• spaces between opposed pedicles along both sides for the passage of spinal nerves exiting the cord: **intervertebral foramina**.

INTERVERTEBRAL DISC

An **intervertebral disc** may be found between each pair of vertebrae. Discs consist of concentric layers of fibrocartilaginous tissue, termed the **annulus fibrosus**, enclosing a jelly-like core—the **nucleus pulposus**. They function as shock absorbers and give flexibility to the column, and also act as spacers or shims so that one vertebra can bend over another without touching its processes.

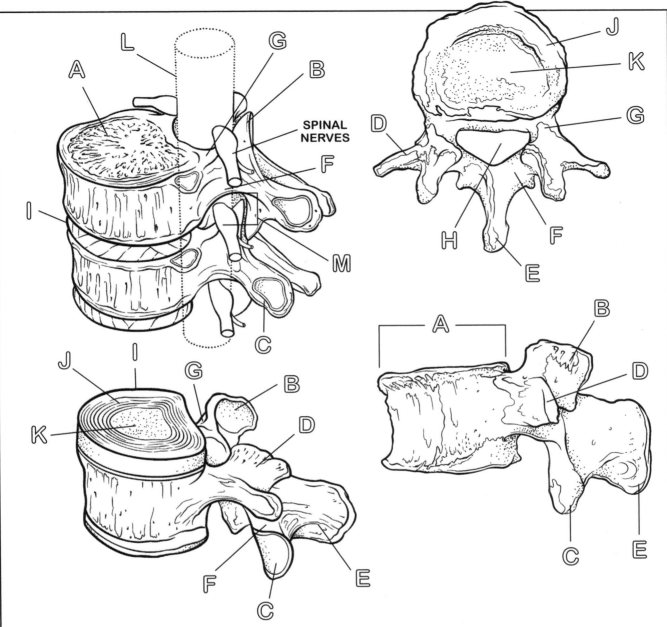

SPINAL NERVES

A VERTEBRAL BODY
B SUPERIOR ARTICULAR PROCESS
C INFERIOR ARTICULAR PROCESS
D TRANSVERSE PROCESS
E SPINOUS PROCESS
F LAMINA

G PEDICLE
H VERTEBRAL FORAMEN
I INTERVERTEBRAL DISC
J ANNULUS FIBROSUS
K NUCLEUS PULPOSUS
L VERTEBRAL CANAL
M INTERVERTEBRAL FORAMINA

VERTEBRAL COLUMN

Together, the thirty-three vertebrae and intervertebral discs form the **vertebral column**, a flexible yet strong support for the trunk. It also protects the spinal cord through a full range of motion. The vertebral column articulates with the skull above and terminates below in a series of tailbones. Just cephalic to the tailbones, a wedge-shaped mass of five fused vertebrae, the **sacrum**, forms a slightly movable articulation with the hip bones in the thorax; the column gives attachment to the ribs.

The vertebral column consists of a collection of some twenty-six separate bones. Note that the bodies of the individual vertebrae are not square; they differ in height between front and back sides, resulting in a series of curves along the length of a column. The group of vertebrae associated with a particular curve look reasonably alike and different intelligibly from groups of vertebrae in other curves. Thus a basis is provided for grouping the vertebrae. In general, each vertebra is progressively larger than its fellow above, culminating in the sacrum, suggesting that more and more weight is borne by the column caudally.

The bodies of the vertebrae are separated from their adjacent fellows by a fibrocartilaginous intervertebral disc. The presence of intervertebral discs implies a certain flexibility in the column as a whole. Happily, stability is not sacrificed for flexibility because, in part, one curve compensates somewhat for the other.

REGIONAL VERTEBRAE

On the basis of shape, size, and other characteristics, vertebrae of the column may be divided into five regions:

• **Cervical vertebrae**, C1 to C7, are the vertebrae of the neck.
• **Thoracic vertebrae**, T1 to T12, are the vertebrae of the thorax and give attachment to the twelve ribs.
• **Lumbar vertebrae**, L1 to L5, support the thorax above and, in effect, function as a pillar, resting solidly on the sacrum below.
• **Sacral vertebrae,** S1 to S5, are fused into one mass and called the sacrum.
• **Coccygeal vertebrae** are comprised of four vertebrae and are collectively called the **coccyx** or **tailbone**.

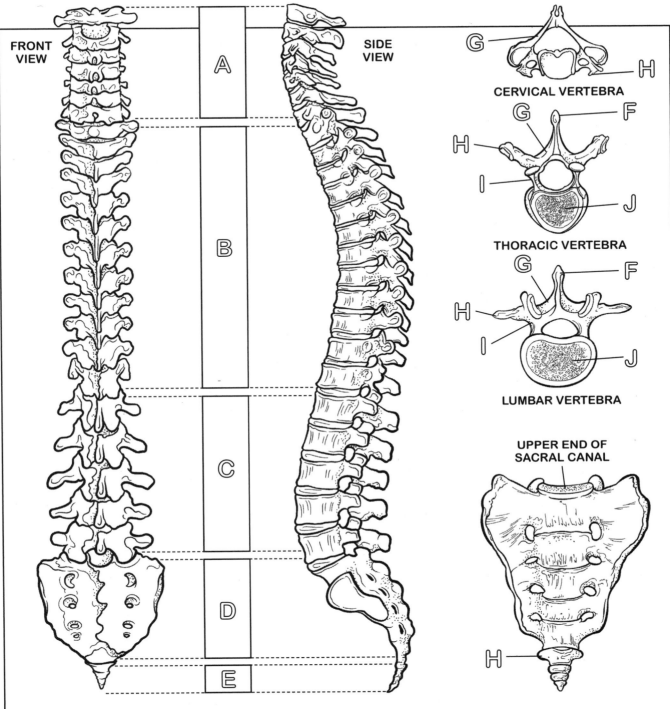

FRONT VIEW

SIDE VIEW

G
H
CERVICAL VERTEBRA

G
F
H
I
J
THORACIC VERTEBRA

G
F
H
I
J
LUMBAR VERTEBRA

UPPER END OF SACRAL CANAL

H

A CERVICAL VERTEBRAE
B THORACIC VERTEBRAE
C LUMBAR VERTEBRAE
D SACRAL VERTEBRAE
E COCCYGEAL
  VERTEBRAE

F SPINOUS PROCESS
G LAMINA
H TRANSVERSE PROCESS
I PEDICLE
J VERTEBRAL BODY

MOVEMENT OF THE VERTEBRAL COLUMN

Movements of the vertebral column are a combination of the *type* of articulation between vertebral arches, the *thickness* of intervertebral discs, and the *functional state* of related ligaments and muscles—as well as the vertebrae and discs themselves.

The movements allowed in the cervical region are **flexion**, **extension**, and a **combined lateral flexion and rotation**. Movement is more limited in the thoracic region, where extension is inhibited by the spinous processes and lateral flexion by the ribs. The lumbar region is as flexible as the cervical region except that the articular processes of adjacent vertebrae make rotation almost impossible. The sacrum is immobile and the coccyx only passively movable.

JOINTS AND LIGAMENTS OF THE VERTEBRAL COLUMN

Each pair of vertebrae is united by three joints: the partly movable joint between the **vertebral body** and the **intervertebral disc**, reinforced by a pair of **longitudinal ligaments** on the **anterior** and **posterior** faces of the vertebral body and disc; and the two joints between **articular processes** of the arches. These are synovial joints of the plane or gliding variety. These joints are reinforced by ligaments joining the vertebral lamina (ligamenta flava), the spine, and the transverse processes, as shown below.

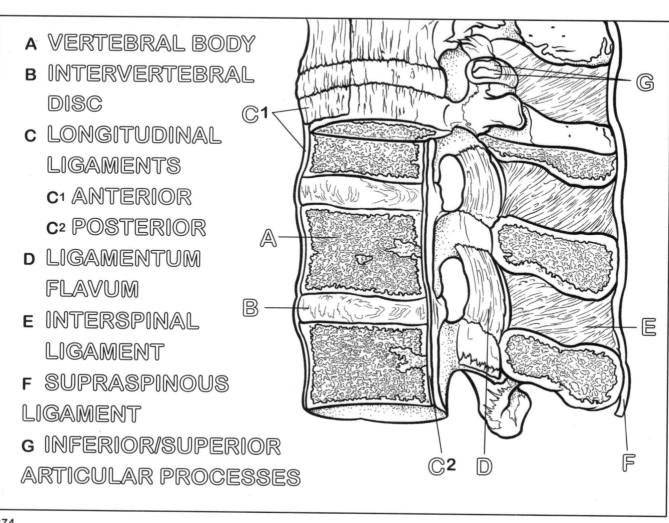

A VERTEBRAL BODY
B INTERVERTEBRAL DISC
C LONGITUDINAL LIGAMENTS
  C1 ANTERIOR
  C2 POSTERIOR
D LIGAMENTUM FLAVUM
E INTERSPINAL LIGAMENT
F SUPRASPINOUS LIGAMENT
G INFERIOR/SUPERIOR ARTICULAR PROCESSES

MUSCLES OF THE BACK

The muscles of the deep back, overlying the posterior aspect of the abdomen and thorax, can be considered in four groups—all of which play an important role in maintenance of erect posture and movements of the head, hip, and trunk including extension, lateral flexion, and rotation. They are the **interspinales**, small muscle bundles between adjacent vertebrae; the **multifidus** and **semispinalis** groups of obliquely oriented muscles, spanning two or more vertebrae running from transverse processes to spines; and the long **erector spinae** muscle group, which may be seen on yourself as muscular bands running parallel to the vertebral column in the lower back. These are the most significant of the deep back muscle groups. A strap muscle bandaging the deep muscles of the neck is termed the splenius.

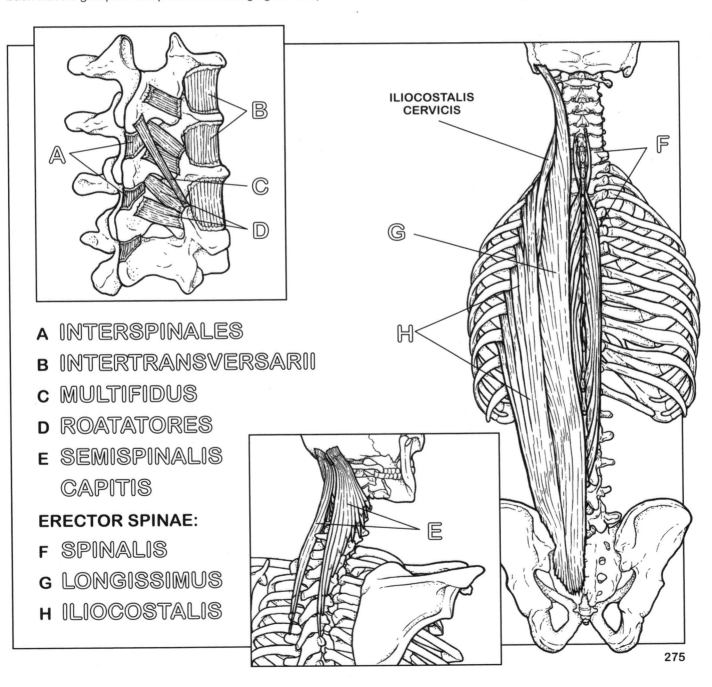

ILIOCOSTALIS CERVICIS

A INTERSPINALES
B INTERTRANSVERSARII
C MULTIFIDUS
D ROATATORES
E SEMISPINALIS CAPITIS

ERECTOR SPINAE:

F SPINALIS
G LONGISSIMUS
H ILIOCOSTALIS

CLINICAL CONSIDERATIONS OF THE SPINE

Curvatures can become affected by disease of individual vertebrae in which there is degeneration of vertebral bone. In response, the body increases the rate of osteogenesis and bony growth (spurs) can overlap intervertebral joints.

Like a tall stack of blocks, the collection of vertebrae with its curves represents a delicately balanced column. Additions to or subtractions from the degree of curvature have immense implications structurally. Ligaments and muscles reinforcing the stability of the column will contract following an alteration, in an attempt to bring the column back in alignment. Sustained contractions may lead to spasmodic contractions and subsequent inflammations, all quite painful.

An abnormal lateral curvature of the vertebral column is termed **scoliosis** (Gr., *scolios*, crooked). Seen infrequently in children, the cause of juvenile scoliosis is unknown (idiopathic). Another kind of scoliosis occurs because of long-standing strain or force acting on one side of the body.

An abnormal thoracic curvature—an exaggerated posterior convexity of the thoracic portion of the column—is termed **kyphosis** (Gr., *kypho*, hump). This condition often leads to cardiopulmonary problems because of the unusual position of the organs with respect to the forces of gravity.

A less serious exaggerated curvature, **lordosis**, involves the lumbar vertebrae and is frequently seen transiently in pregnant women who compensate for the muscular strain of the added anterior weight by taking on a "proud stance." Exercises following parturition usually reduce this exaggerated curve.

ABNORMAL CURVATURES

A SCOLIOSIS

- Uneven shoulders
- One shoulder blade appears more prominent than the other
- Uneven waist
- One hip higher than the other

B KYPHOSIS

- Forward posture of the head
- Upper back appears hunched over
- Difference in shoulder height
- Tight hamstrings

C LORDOSIS

- Exaggerated inward curve of the lower spine
- Buttocks appear pronounced
- Arch between lower back and a flat surface when lying down

THE THORACIC WALL: STERNUM & RIBS

The **thorax** is a cage for the lungs, heart, and other significant structures. It is more or less open into the neck above at the **superior aperture** and closed below at the **inferior aperture** by the dome-like thoracic diaphragm, as we shall see.

Place your hand behind your back at the level of the lower part of the scapula—as if you were scratching your back—to feel the spines of the thoracic vertebrae.

STERNUM

Most boundaries of the sternum can be felt and palpated.

Start above, where you feel the concave, crescent-shaped upper border, the **jugular notch** (at about T2). Spreading your fingers laterally from the notch, palpate the **clavicles**. Immediately deep and inferior to the **sternoclavicular joint**, the **first costal cartilage** in the short space becoming the first rib starts its journey around to the **first thoracic vertebra**. The circle of bone thus created, which cannot be felt, constitutes the **superior aperture**.

Now feel down two or three fingerbreadths below the jugular notch for a slight bony bump or rise signaling the **sternal angle**—a synovial joint between the **manubrium** and the **body** of the sternum. The sternal angle represents a landmark for deeper visceral structures in the thorax and is usually on a horizontal plane with T4.

The **sternal body**—at the level of the seventh costal cartilage—articulates with the spear-tip **xiphoid process** (third part of the sternum), palpated only with difficulty.

RIBS

Although there are twelve ribs on each side, no one is exactly like another. Ribs have both bony (costo-) and cartilaginous (chondro-) components. Each bony rib is characterized by a shaft that begins anteriorly with the costochondral joint and arches backward and laterally to a point short of the vertebra. It then turns abruptly forward at the angle of the rib, narrows to form a neck, and terminates at the vertebral articulation as the head.

The costal cartilages are of the hyaline variety and join the bony rib to the sternum anteriorly.

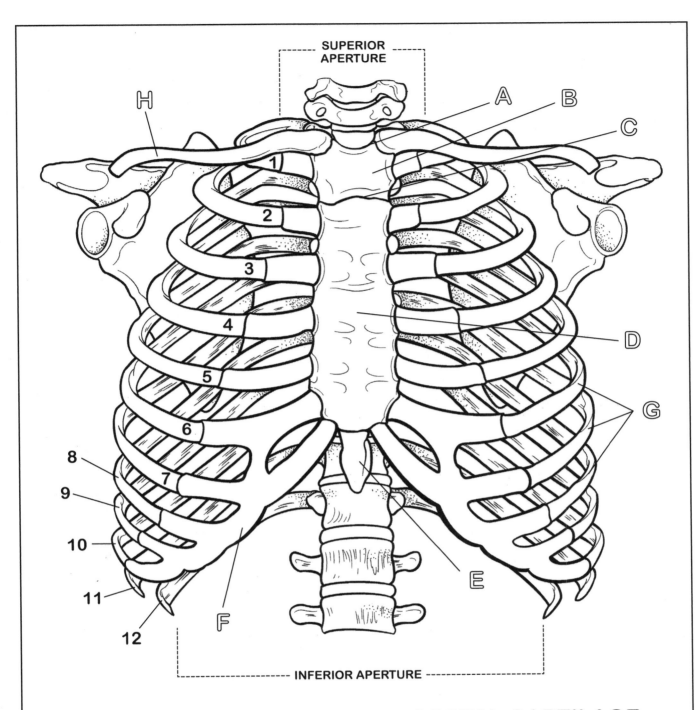

SUPERIOR APERTURE

INFERIOR APERTURE

STERNUM:

A JUGULAR NOTCH

B MANUBRIUM

C STERNAL ANGLE

D STERNAL BODY

E XIPHOID PROCESS

F COSTAL CARTILAGE

G RIBS

H CLAVICLE

INTERCOSTAL MUSCLES
& THE DIAPHRAGM

The thoracic cavity is a closed cavity—not in any way in communication with the outside. Its walls are movable and therefore can change the volume of the cavity. (The elastic lungs within the thorax are open to the outside, of course, via the conducting airways of the trachea, larynx, pharynx, and nasal and oral cavities.)

INTERCOSTAL MUSCLES

The space between adjacent pairs of ribs is called an **intercostal space**. The musculature of each of the eleven intercostal spaces is oriented in three incomplete layers: the **external intercostal** musculature arises from the lower border of the rib above and descends downward and medially to insert on the upper border of the rib below. The **internal intercostals** have similar attachments except that the fibers are oriented 90° with respect to the external intercostals. Finally, the **innermost intercostals** are composed of discontinuous sheets of muscles spanning one or more intercostal spaces—and again, these fibers are oriented 90° with respect to the internal intercostals.

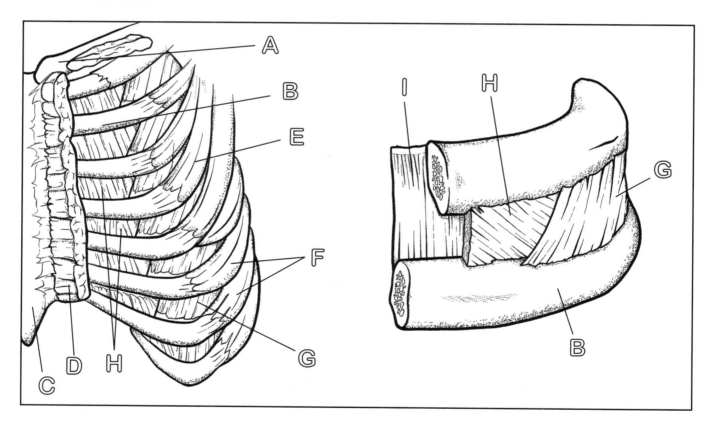

A CLAVICLE

B RIBS

C STERNUM

D PECTORALIS MAJOR **(CUT)**

E PECTORALIS MINOR

F SERRATUS ANTERIOR

G EXTERNAL INTERCOSTAL

H INTERNAL INTERCOSTAL

I INNERMOST INTERCOSTAL

DIAPHRAGM

The diaphragm is a musculotendinous structure between the abdomen and the thorax. The muscular fibers of the diaphragm originate from the ribs, the xiphoid process, and vertebrae T12 to L3, that is, from all sides around the thoracoabdominal opening. The muscle fibers project inward toward the **central tendon**, which is punctured by three openings or **hiatuses**. On the right side, the **inferior vena cava** (draining the lower limbs, pelvis, and the body wall) passes through. Centrally, just in front of the vertebral bodies, is the **aorta**, while slightly to the left, the **esophagus** passes through to become the stomach below. Since the diaphragm arches upward centrally, the vena cava passes through the middle of the diaphragm at a higher vertebral level (T8) than the more peripheral esophagus (T10) or aorta (T12).

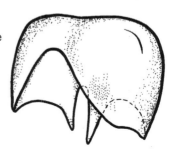

The intercostal musculature and the diaphragm act to increase intrathoracic dimensions so **inspiration** (taking air into the lungs) can occur. The act of **expiration** (taking air out of the lungs) normally requires no muscular assistance since the natural elasticity of the lungs and thoracic wall brings about the decreases in lung and thoracic wall dimensions. The **crus** of the diaphragm are two tendinous structures, extending below the diaphragm to the vertebral column, forming a tether for muscular contraction.

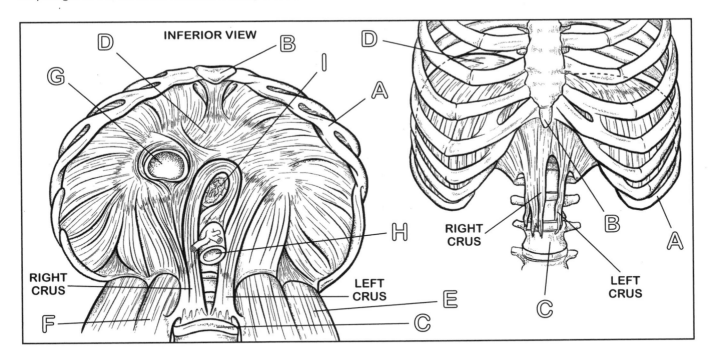

A RIBS

B XIPHOID PROCESS

C LUMBAR SPINE

D CENTRAL TENDON

E QUADRATUS LUMBORUM

F PSOAS MAJOR MUSCLE

HIATUSES:

G INFERIOR VENA CAVA

H AORTA

I ESOPHAGUS

BOUNDARIES OF
THE ABDOMINAL WALL

Because of their muscular configuration, the walls of the abdominal cavity are described as being *anterolateral* and *posterior*. The orientation of the muscles in the anterolateral walls is quite reminiscent of the muscles and thoracic wall, suggesting some degree of continuity in development.

BOUNDARIES OF THE ANTEROLATERAL ABDOMINAL WALL
Unlike the thoracic wall, there is no bony framework to the major portion of the abdominal wall. You can feel this on yourself: Palpate the margin of the ribs from the xiphoid process laterally. Below these bony structures there is only muscle, fascia, and skin protecting the fragile abdominal viscera from the external environment! At the level of the umbilicus, feel latterly for the crests of the ilium and feel for the sharp drop-off anteriorly to the pubic bones. These bony landmarks not only indicate to you the boundaries of the anterolateral abdominal wall but they are also the points of origin and insertion of certain important muscles.

EXTERNAL OBLIQUE
The **external oblique** is the most superficial of the three anterolateral muscles, and its fiber orientation is like that of the external intercostals. In lean individuals, the fibers of origin of the external oblique may be seen descending obliquely from the ribs below the origin of the **serratus anterior**. The left and right fibers of the external oblique form a massive aponeurosis anteriorly and interdigitate in the midline as part of the **linea alba**. Inferiorly, the muscle ends as a free border, turned or curled upon itself to form a ligament between anterior superior iliac spine and pubis. Can you palpate the inguinal ligament on yourself?

INTERNAL OBLIQUE
Deep to external oblique, its fibers oriented somewhat like those of the internal intercostals, one finds the **internal oblique**. It is, of course, not palpable, and its attachments are noted in the adjacent illustration.

TRANSVERSUS ABDOMINIS
Deep to the internal oblique, the **transversus abdominis** reinforces the muscular anterolateral wall. Like its more superficial fellows, the transversus abdominis contributes to the sheath of the **rectus abdominis** and to the inguinal canal as well.

RECTUS ABDOMINIS AND ITS SHEATH
The straight muscle of the abdomen is easily seen in well-developed, lean males and females. The lateral margin of the muscle creates a skin shadow on the abdominal wall as well. The borders and extent of the rectus abdominis can probably be ascertained on yourself. The fascial **rectus sheath** is reinforced by the interlacing aponeuroses of the three anterolateral muscles just described.

The three anterolateral muscles generally act together in flexing the trunk. The rectus abdominis helps out only when there is considerable flexion, as in a sit-up exercise. The anterolateral muscles (except the rectus) are active in tensing the abdominal wall as in straining and coughing and are of primary importance in forced expiration.

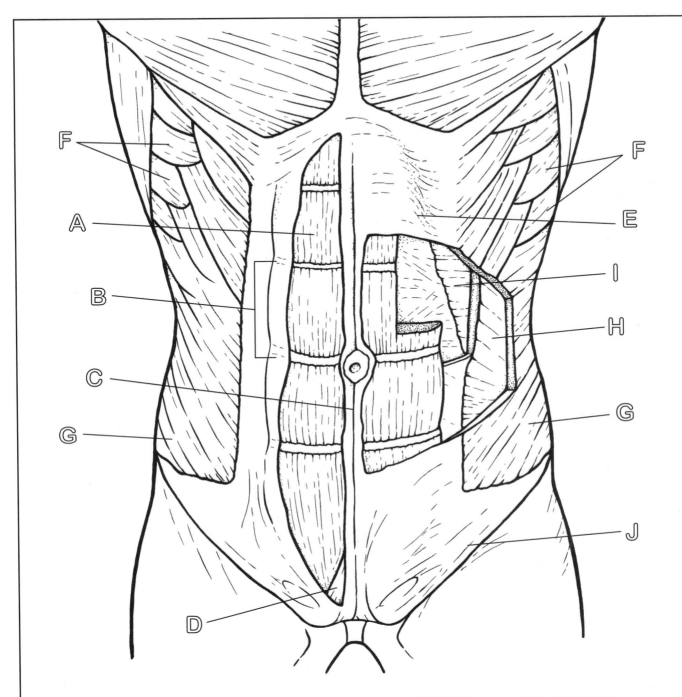

A RECTUS ABDOMINUS

B LINEA SEMILUNARIS

C LINEA ALBA

D PYRAMIDALIS

E RECTUS SHEATH

F SERRATUS ANTERIOR

G EXTERNAL OBLIQUE

H INTERNAL OBLIQUE

I TRANSVERSUS ABDOMINUS

J INGUINAL LIGAMENT

POSTERIOR
ABDOMINAL WALL

The muscles of the anterolateral abdominal wall are innervated by anterior rami of spinal nerves T7 to L1, which wrap around the wall not unlike the intercostal nerves wrap around the thoracic wall. Thus you might suspect that since the intercostal nerves pass between the second and third layers of intercostal musculature, the nerves to the abdominal wall must pass between internal oblique and transversus abdominis, and they indeed do, giving off branches as they go. The cutaneous distribution to the abdominal wall is oriented in the same fashion as that to the thoracic wall.

The cutaneous distribution of nerves to the abdomen and thorax is important to the physician, for often visceral pain will be referred to cutaneous regions that have the same spinal cord level of innervation. Thus, an itching or painful sensation of the skin of the abdomen at the level of the navel may be a reflection of an inflammation of the lining of the lungs (pleurisy) at the level supplied by the tenth thoracic nerve.

INGUINAL REGION
The inguinal region (not pictured) is a portion of the lower quadrants of the anterior abdominal wall characterized by a fibromuscular canal transmitting the spermatic cord (in males) from the testes in the scrotum.

The spermatic cord consists of blood vessels and a duct. The duct transports sperm cells from the testis (where they are produced) through the abdominal wall, by way of the inguinal canal, to the urethra in the prostate gland. The urethral duct passes through the penis and is, of course, open to the outside. In this way sperm cells leave the body of the male.

POSTERIOR ABDOMINAL WALL
Shown opposite is the posterior wall of the abdomen, internal to the muscles of the back and anterior to the transverse processes of the lumbar vertebrae. It consists of five major structures:
• Five **lumbar vertebrae**.
• The quadrate-shaped muscle of the lumbar region (**quadratus lumborum**) located between the twelfth rib and the iliac crest.
• The **psoas muscle** (a flexor of the thigh and vertebral column).
• The **iliacus muscle** (a flexor of the thigh and vertebral column).
• The **diaphragm** (in part).

The significance of this wall lies in the fact that between this wall and the peritoneum lining are most of the nerves and vessels supplying the abdomen, not to mention the **kidneys, adrenal glands**, and portions of the large intestine. Some of these are:
• The abdominal aorta.
• The **inferior vena cava**.
• The autonomic plexuses on the aorta, supplying sympathetic and parasympathetic innervation to the abdominal viscera.
• The retroperitoneal (behind the peritoneum) viscera.
• The lumbar plexus of somatic nerves.

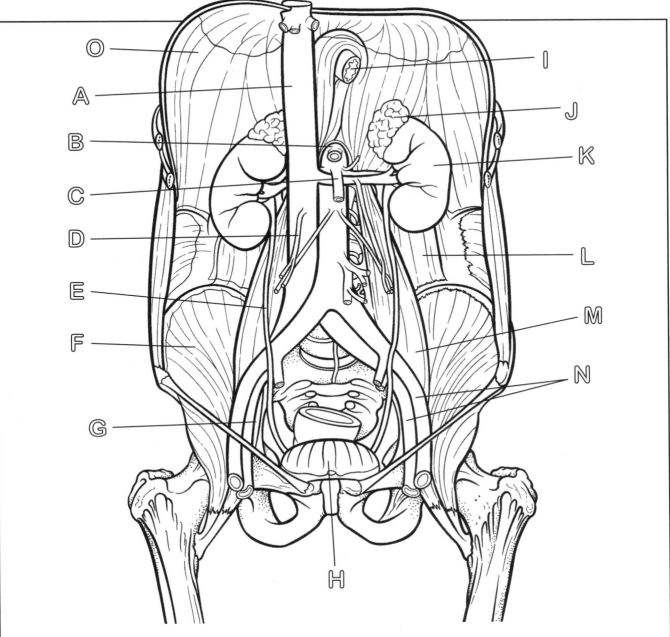

A INFERIOR VENA CAVA

B CELIAC TRUNK

C SUPERIOR MESENTERIC ARTERY

D GONADAL VEIN

E URETER

F ILIACUS MUSCLE

G FEMORAL NERVE

H URINARY BLADDER

I ESOPHAGUS

J LEFT ADRENAL GLAND

K LEFT KIDNEY

L QUADRATUS LUMBORUM

M PSOAS MAJOR

N ILIAC ARTERY & VEIN

O DIAPHRAGM

SECTION NINE:
THORACIC VISCERA

We have learned of a cage created from slender, curved bones—ribs—attached behind to the twelve thoracic vertebrae and in front to the sternum. In the space between each pair of ribs is a sheet of intercostal muscle, three layers deep. But what about the viscera of the chest sheltered within?

The heart and the lungs are the principal residents of the thorax. It is an efficient housing for both; for the heart because it can pump oxygen (within the medium of blood) a short distance to the head and upper limbs, overcoming the force of gravity without undue strain; for the lungs because they rest on the diaphragm, the floor of the thorax, which performs as a muscular bellows, drawing air through the lungs and airway system.

Anatomically, the organization of structures within the thorax is quite simple: The lungs, their lining, and their vessels and nerves occupy the outer two thirds of the cavity; the heart and transient structures, lying deep to the sternum, fill the region between the lungs, called the mediastinum.

Thoracic viscera also includes other organs of the respiratory, cardiovascular, and lymphatic systems; the middle and lower airways; the inferior portion of the esophagus; the great arteries bringing blood from the heart out into general circulation, and the major veins into which the blood is collected for transport back to the heart; and the thymus gland, all protected by the bony rib cage.

THORACIC VISCERA

The **mediastinum** is a division of the thoracic cavity; it contains the heart, thymus gland, portions of the esophagus and trachea, and other structures. The mediastinum is arranged into four compartments: **anterior, posterior, middle,** and **superior**. The upper limit of the mediastinum is set by the superior aperture of the **thorax**, a thoroughfare between neck and chest. The remaining boundaries are the **diaphragm** below, the medial surface of the **lungs** laterally, the **sternum** anteriorly, and posteriorly, the bodies of **thoracic vertebrae**.

From the illustrations, note:
• The principal resident of the mediastinum is the **heart**, occupying the middle compartment.
• The anterior mediastinum is virtually free of major structures.
• The posterior compartment is largely filled with structures en route to the lungs (**trachea, bronchi**), and abdomen (**aorta, esophagus, nerves**, etc.).
• The superior compartment contains structures in transit from the neck to the posterior mediastinum and points below (esophagus, etc.), the great vessels springing from and draining into the heart, as well as local viscera such as the **thymus gland**.

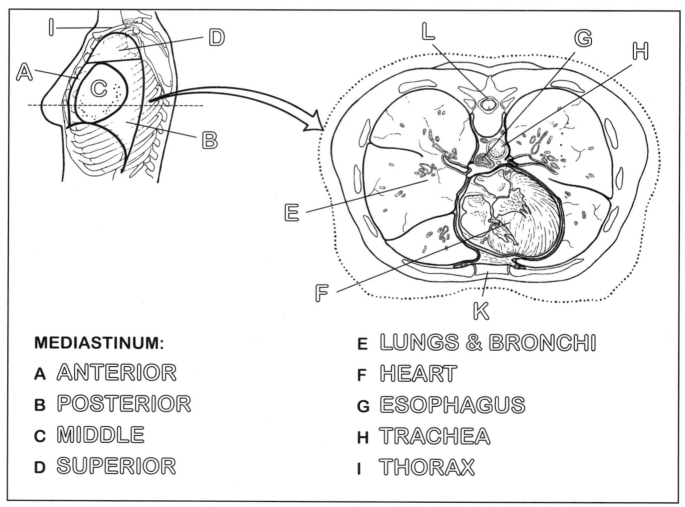

MEDIASTINUM:
A ANTERIOR
B POSTERIOR
C MIDDLE
D SUPERIOR

E LUNGS & BRONCHI
F HEART
G ESOPHAGUS
H TRACHEA
I THORAX

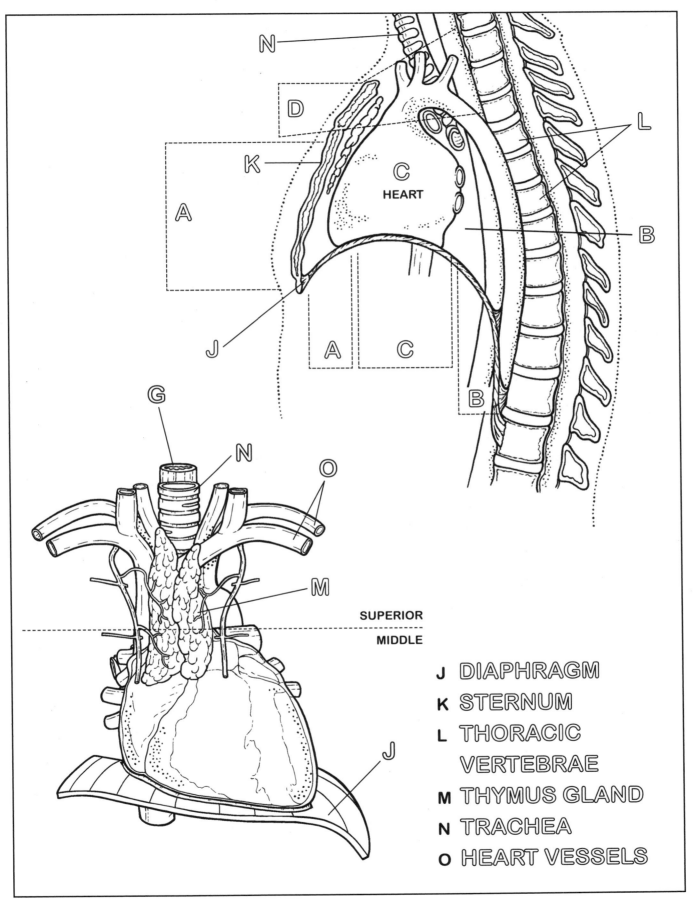

N

D

K

A

C
HEART

J

L

B

A C

B

G

N

O

M

SUPERIOR

MIDDLE

J

J DIAPHRAGM
K STERNUM
L THORACIC
VERTEBRAE
M THYMUS GLAND
N TRACHEA
O HEART VESSELS

THE HEART

The heart is a muscular pump that provides the force for the movement of blood through all the vessels within the body. Thanks to the heart, the tissues of the body receive oxygen and nutrition, and have a disposal service for discharge of carbon dioxide and other "undesirable" elements. Because of the heart and circulatory system, tissues of the body can release secretory material that within seconds exerts an influence some distance from its source. Medications injected into the body are distributed to all the "nooks and crannies" and the body's defensive weaponry—antibodies and phagocytes—may be mobilized to areas of infection and inflammation.

Conversely, diminished heart action can result in insufficient tissue oxygenation, venous pooling (stasis), filling of tissue spaces (edema) due to venous back pressure, and other conditions. Arrested heart action causes death because the cells of the brain initially, and all cells subsequently, die when the oxygen supply stops.

As is the case with any pump, there is an input as well as an output side. An input chamber of the heart is called an **atrium**; an output chamber, a **ventricle**. The apparent simplicity of this operation is complicated by the fact that two pumps are necessary:

Pump number 1 (the right side) collects deoxygenated blood from the body tissues via two great veins (input) and drives it into the lungs for aeration by way of one great arterial trunk (output). Pump number 2 (the left side) collects freshly oxygenated blood from the lungs by means of four veins (input) and drives it to the body's tissues through one great artery (output).

These two pumps, composed of two atria and two ventricles, in a triple-layered bag (the **pericardium**), punctured by orifices for six large veins and two great arteries, constitute the heart.

FLOW CIRCUITS
In the body there are two circuits of blood flow. One involves the right atrium and ventricle, and the other the left atrium and ventricle. The **pulmonary circuit** is: right ventricle (pump) → lungs → left atrium (reservoir); and the **systemic circuit** is: left ventricle (pump) → body tissues → right atrium (reservoir). The pulmonary circuit functions to oxygenate blood returning to the heart from the body tissues. The systemic circuit functions to pump oxygenated blood to the tissues of the body and return it, deoxygenated, to the heart.

Deoxygenated blood enters the right atrium via the inferior and superior venae cavae (IVC, SVC). The IVC generally drains the lower extremities and the abdomen, while the SVC generally drains the upper extremities: the head, neck, and thorax.

PERICARDIUM
The heart is enveloped by pericardium, a double-walled sac. The inner wall, or **visceral pericardium**, adheres closely to the **myocardium** (or heart muscle). At the base of the heart the inner wall turns outward and heads back to become the inner surface of the **parietal pericardium**. The space between these two pericardial layers constitutes the **pericardial cavity**.

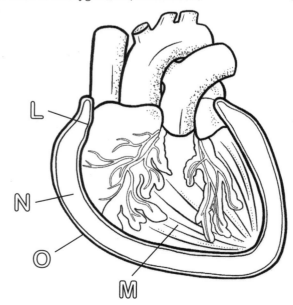

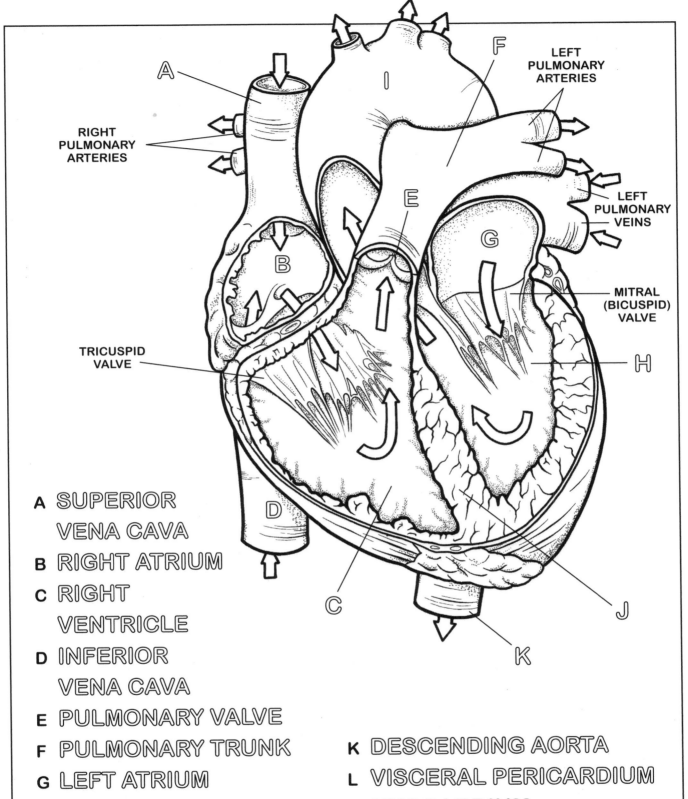

RIGHT
PULMONARY
ARTERIES

A

LEFT
PULMONARY
ARTERIES

F

I

LEFT
PULMONARY
VEINS

E

G

MITRAL
(BICUSPID)
VALVE

B

H

TRICUSPID
VALVE

J

D

J

C

K

A **SUPERIOR VENA CAVA**

B **RIGHT ATRIUM**

C **RIGHT VENTRICLE**

D **INFERIOR VENA CAVA**

E **PULMONARY VALVE**

F **PULMONARY TRUNK**

G **LEFT ATRIUM**

H **LEFT VENTRICLE**

I **AORTIC ARCH**

J **SEPTUM**

K **DESCENDING AORTA**

L **VISCERAL PERICARDIUM**

M **MYOCARDIUM**

N **PERICARDIAL CAVITY**

O **PARIETAL PERICARDIUM**

RESPIRATORY STRUCTURES OF THE MEDIASTINUM

THE TRACHEA AND BRONCHI

The trachea, also known as the "windpipe," is the distal continuation of the larynx. It passes through the superior mediastinum and into the posterior mediastinum, where it terminates by bifurcating into the right and left primary bronchi, each of which supply a lung. Its upper border is at the level of the C6 vertebrae; the bifurcation occurs at about T4.

The supporting framework of the trachea and bronchi is **hyaline cartilage** in the form of horseshoe-shaped rings with fibrous interspaces of **annular ligaments**. As one proceeds along the bronchi distally from the **bifurcation**, the rather regular arrangement of rings can be seen to fragment into irregular plates of **bronchial cartilage**. Within the trachea and primary bronchi is a mucosa lined with pseudostratified columnar epithelium, including goblet (mucous) cells and cilia. Mucous glands heavily populate the underlying connective tissue.

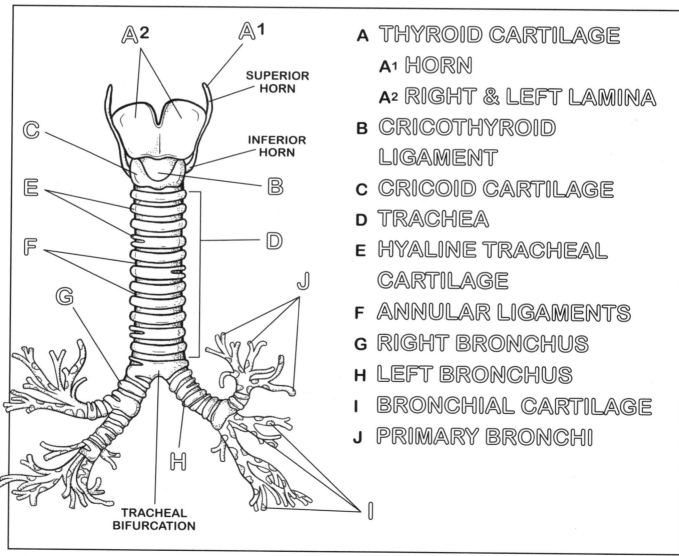

A THYROID CARTILAGE
A1 HORN
A2 RIGHT & LEFT LAMINA
B CRICOTHYROID LIGAMENT
C CRICOID CARTILAGE
D TRACHEA
E HYALINE TRACHEAL CARTILAGE
F ANNULAR LIGAMENTS
G RIGHT BRONCHUS
H LEFT BRONCHUS
I BRONCHIAL CARTILAGE
J PRIMARY BRONCHI

SUPERIOR HORN

INFERIOR HORN

TRACHEAL BIFURCATION

VAGUS NERVES AND BRANCHES

The parasympathetic nerves responsible for the regulation of internal organ functions (digestion, heart rate, and respiratory rate) are the **vagus nerves**. Originating in the brainstem, vagi are distributed to the regions of the pharynx and larynx, the heart, lungs, and abdominal viscera. The vagi pass into the superior mediastinum in company with the common carotid arteries and internal jugular veins. While in the neck they give off the cardiac branches, forming a plexus at the base of the heart with cardiac branches from the sympathetic chain. Vagi also give off branches to the bronchi and lungs (**pulmonary nerves**), the larynx and trachea (**recurrent laryngeal nerves**), and the **esophageal plexus**.

PHRENIC NERVES

The **phrenic nerve** is a bilateral, mixed nerve that originates from the cervical nerves in the neck and descends through the thorax to innervate the diaphragm. It is the only source of motor innervation to the diaphragm and therefore plays a crucial role in breathing; sever it and one is left with a lethal break in the circuit between the respiratory center in the medulla and the origins of phrenic and intercostal nerves.

The phrenic nerves' roots (third, fourth, and fifth cervical spinal nerves) are also roots for the sensory nerves to the neck and shoulder. Thus, irritation of the membranes of the lung adjacent to the diaphragm may show up in pain of the anterior neck and shoulder—an example of **referred pain**.

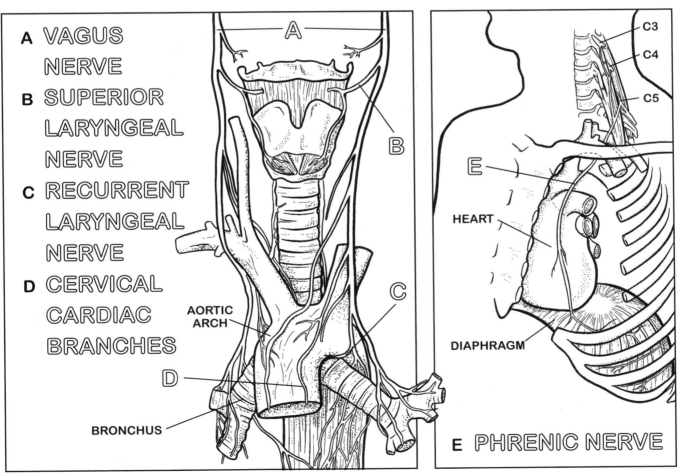

A VAGUS NERVE
B SUPERIOR LARYNGEAL NERVE
C RECURRENT LARYNGEAL NERVE
D CERVICAL CARDIAC BRANCHES

AORTIC ARCH
BRONCHUS

C3
C4
C5

HEART
DIAPHRAGM

E PHRENIC NERVE

DIGESTIVE ORGANS: ESOPHAGUS, STOMACH, INTESTINES

Note: The esophagus resides in the thoracic region (viscera) and the stomach and intestines reside in the abdominal region (viscera), but for ease of understanding the digestive tract, they are all discussed here.

The **esophagus** is the distal extension of the pharynx and proceeds through the superior and posterior mediastinum, posterior to the trachea and anterior to the aorta, to pass through the diaphragm, dilate, and become the stomach. It is a collapsed fibromuscular tube that incorporates stratified squamous epithelium in a mucous lining consistent with its function. The underlying connective tissue contains glands whose secretions (as well as those of the oral cavity and pharynx above) facilitate passage of food material. The undulating action of the smooth muscle (peristalsis) moves food (bolus) swiftly through the esophagus. These transitions mirror the fact that food ingested into the esophagus passes from a stage of processing that is *voluntarily* initiated, i.e., chewing and swallowing, to a stage that is *involuntary*, i.e., passage through the esophagus. In terms of digestion, the esophagus plays no significant role.

The **stomach** (L., *gaster*) is the tubular receptacle of the gastrointestinal tract. It is the dilated continuation of the esophagus, yet differs significantly in its microscopic make-up. Elastically hinged by the greater and lesser omenta, the stomach has a great deal of latitude in its movement—its shape is subject to a great deal of variation from gastric contents, position of the body, position of neighboring viscera, etc. When full, the stomach may droop into the pelvis. The two extremities of the stomach are fairly well fixed—above by the diaphragm at about the tenth thoracic vertebral level; and below by the **duodenum**, which is for the most part retroperitoneal.

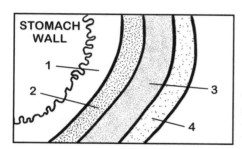

The stomach consists of three parts: the **fundus** above, the **pyloric** portion below, and the **body** in between; and has two curvatures: the convex **greater curvature** and the concave **lesser curvature**. The stomach wall consists of four layers, shown left: an inner **mucosa (1)**, an underlying **submucosa (2)**, a muscular **tunic (3)**, and an outer coat of peritoneum, **serosa (4)**. The interior surface of the stomach is characterized by longitudinal folds (**rugae**) which serve to increase the secretory and absorptive capacity of the stomach. This interior surface is coated with mucous.

The **small intestine** consists of several feet of mobile, actively metabolic digestive tubing, highly coiled within the framework of the large intestine. The small intestine has three parts: the C-shaped **duodenum**, the **jejunum**, and the **ileum**.

The **large intestine** consists of **cecum, colon, rectum,** and **anal canal**. Roughly five feet in length, it begins in the lower right quadrant of the abdomen where the ileum joins the cecum. The cecum is a sac-like affair containing three orifices—one from the ileum, one to the appendix, and one to the ascending colon. The appendix is a blind tube arising from the cecum and has no known digestive function.

The retroperitoneal ascending colon passes upward along the right side of the abdominal cavity, bumps into the underside of the liver, then turns medially to become the transverse colon suspended by its own mesentery, the transverse mesocolon. Drooping in the midline, the transverse colon ascends as it approaches the spleen, then abruptly turns downward to become retroperitoneal as the descending colon. At the pelvic rim, the colon makes a sort of S- (sigmoid) turn to drop into the pelvis. Hinged by mesentery (sigmoid mesocolon), the sigmoid colon becomes the rectum in the pelvis. Passing through and bound by the levator ani muscle, the rectum merges into the anal canal with its distal orifice, the anus.

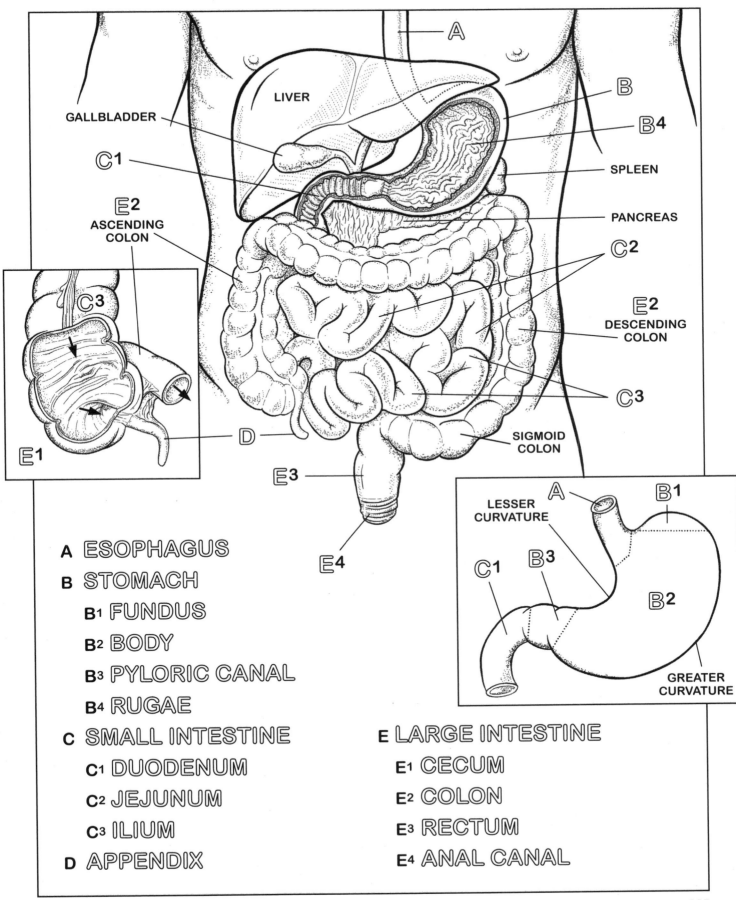

GALLBLADDER

LIVER

A

B

B4

SPLEEN

PANCREAS

C1

E2
ASCENDING
COLON

C3

C2

E2
DESCENDING
COLON

C3

D

E1

SIGMOID
COLON

E3

E4

LESSER
CURVATURE

A

B1

C1

B3

B2

GREATER
CURVATURE

A ESOPHAGUS

B STOMACH

 B1 FUNDUS

 B2 BODY

 B3 PYLORIC CANAL

 B4 RUGAE

C SMALL INTESTINE

 C1 DUODENUM

 C2 JEJUNUM

 C3 ILIUM

D APPENDIX

E LARGE INTESTINE

 E1 CECUM

 E2 COLON

 E3 RECTUM

 E4 ANAL CANAL

MISCELLANEOUS STRUCTURES OF THE MEDIASTINUM
AZYGOS VEIN, THORACIC DUCT & THYMUS GLAND

AZYGOS VEIN

Lying on the vertebral bodies along the posterior thoracic wall, the **azygos vein** and its tributaries drain the posterior intercostal spaces and the posterior abdominal wall. Remember that the **inferior vena cava** (IVC) enters the right atrium of the heart immediately on passing through the diaphragm and so is not available for venous drainage in the chest; the azygos system of veins has that chore.

The azygos vein empties into the **superior vena cava,** and therein lies its special significance: Should the IVC or one of the major tributaries become blocked, or should the hepatic portal vein draining the intestinal tract become obstructed, venous blood can find its way to the heart by means of the azygos system.

THORACIC DUCT

The **thoracic duct** is the principal vessel of the lymphatic system. It lies in the body of the thoracic vertebrae and is deep to the thoracic aorta. It drains the lymphatic trunks of the lower limbs, pelvis, and abdomen; it drains the lymph nodes of the right chest, neck, and upper limbs; and it empties into the junction of the left subclavian and internal jugular veins.

Upward of three quarts of lymph fluid pass through the duct each day. Occlusion of the thoracic duct allows excess fluids to accumulate in the body spaces (edema). It is apparent, then, that the vascular system cannot handle the draining of body tissues without these lymphatic vessels.

THYMUS GLAND

The **thymus gland** lies in the anterior mediastinum atop the pericardium at the base of the **heart**. It is just deep to the manubrium of the sternum. The thymus is a highly complex gland which is most active in children. It tends to decrease in size and activity after childhood.

The thymus produces lymphocytes. It is thought by some investigators to "seed" the body with lymphatic tissue cells (lymphocytes, plasma cells, etc.) which in some unexplained manner provide the body with the competence to resist the invasion of foreign bodies, such as bacteria and viruses. It is part of the body's immunological apparatus.

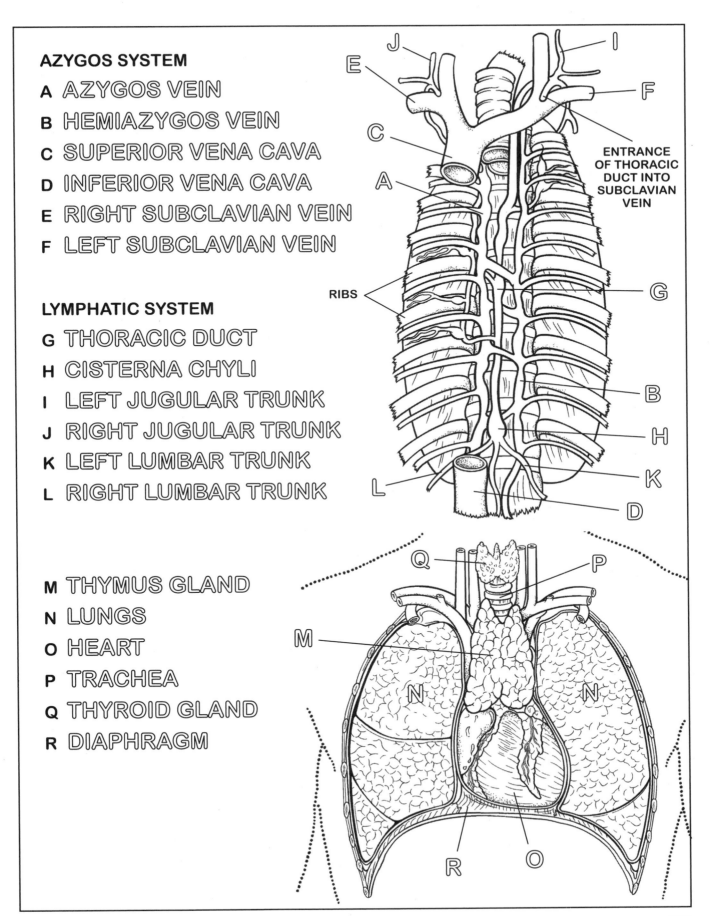

AZYGOS SYSTEM

A AZYGOS VEIN
B HEMIAZYGOS VEIN
C SUPERIOR VENA CAVA
D INFERIOR VENA CAVA
E RIGHT SUBCLAVIAN VEIN
F LEFT SUBCLAVIAN VEIN

LYMPHATIC SYSTEM

G THORACIC DUCT
H CISTERNA CHYLI
I LEFT JUGULAR TRUNK
J RIGHT JUGULAR TRUNK
K LEFT LUMBAR TRUNK
L RIGHT LUMBAR TRUNK

M THYMUS GLAND
N LUNGS
O HEART
P TRACHEA
Q THYROID GLAND
R DIAPHRAGM

ENTRANCE
OF THORACIC
DUCT INTO
SUBCLAVIAN
VEIN

RIBS

LUNGS & PLEURAE

The lungs occupy the greatest portion of the thorax and represent that part of the body in which the internal vascular system comes closest to contacting the external environment (i.e., the atmosphere). Indeed, the blood-air cell interfaces of the lung are so thin that during exercise almost a quart of oxygen can diffuse into the blood per minute. In terms of volume, 60 percent of the lung is blood. Every minute blood flows through the vessels of the lungs, over a gallon of air is brought into contact with it.

GROSS STRUCTURE

The lungs, resembling light and frothy sponges, fill up the lateral two thirds of the chest cavity and are separated from one another by the mediastinum. Characterized by a sharp anterior border, a rounded posterior border, a convex lateral surface, and a concave medial surface, lungs are softly pyramid-shaped, with a base resting on the **diaphragm** and the **apex** projecting up through the superior thoracic aperture into the root of the neck. Each lung is anchored firmly to the mediastinum at the root (**hilus**) by the **bronchi** and **pulmonary vessels**. The right lung has **upper**, **middle**, and **lower lobes** while the left lung has an **upper** and **lower lobe**, along with a **cardiac notch** to accommodate the heart. **Oblique** and **horizontal fissures** demarcate the lobes. Being air-filled and malleable, the lungs contain multiple notches and grooves for the aorta, ribs, azygos vein, and other neighboring structures.

PLEURAE

The **pleurae** are the membranes of the lung. These layers are continuous with one another, creating a sleeve-like pulmonary ligament at the **root** of the lung. The **visceral pleura** clothes the lung tissue while the **parietal pleura** is pressed against the inner **thoracic wall** and forms the sides of the mediastinum. The space between the two layers constitutes the **pleural cavity**. The pleural membranes are mesothelial in character, secreting a film of serous fluid which eliminates friction between them and provides a degree of surface tension (adherence) between them. When the pleurae become inflamed and their serous secretions are interrupted, a painful condition known as **pleurisy** develops.

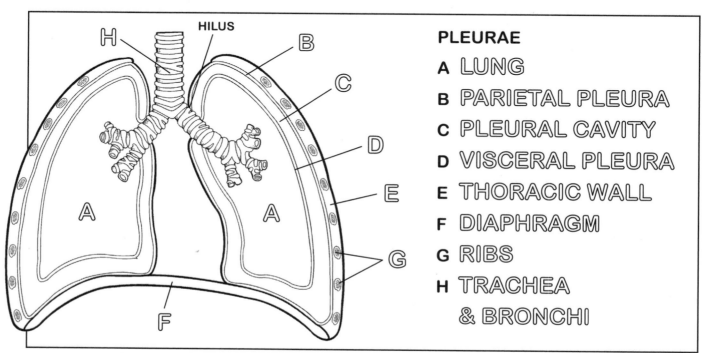

HILUS

PLEURAE

A LUNG
B PARIETAL PLEURA
C PLEURAL CAVITY
D VISCERAL PLEURA
E THORACIC WALL
F DIAPHRAGM
G RIBS
H TRACHEA & BRONCHI

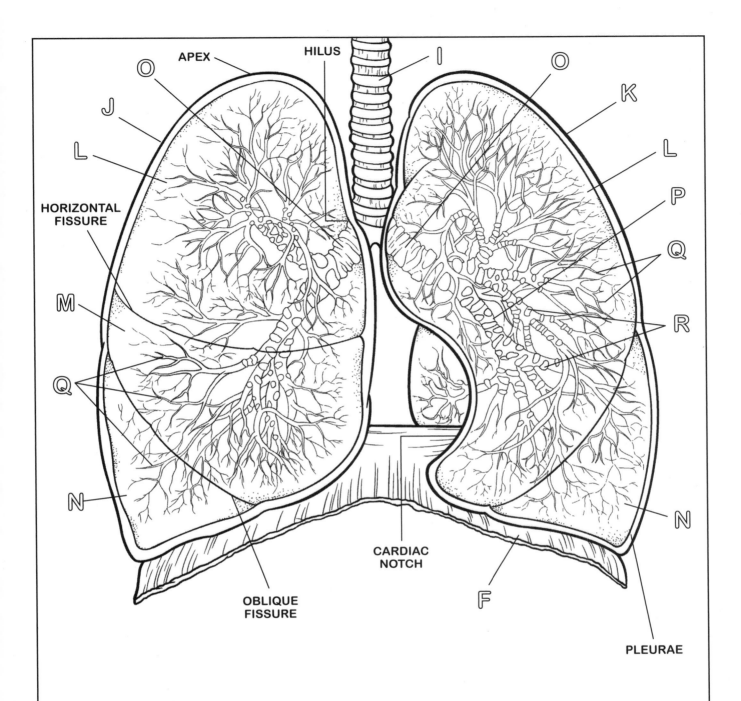

APEX

HILUS

HORIZONTAL
FISSURE

CARDIAC
NOTCH

OBLIQUE
FISSURE

PLEURAE

I	TRACHEA	N	LOWER LOBE
J	RIGHT LUNG	O	PRIMARY BRONCHI
K	LEFT LUNG	P	SECONDARY BRONCHI
L	UPPER LOBE	Q	TERMINAL BRONCHIOLES
M	MIDDLE LOBE	R	SEGMENTAL BRONCHUS

LUNGS: ARTERIES, VEINS, INNERVATION & GASEOUS EXCHANGE

BLOOD AND NERVE SUPPLY TO THE LUNGS

The principal source of oxygenated blood to lung tissue are the bronchial arteries, which spring from the aorta. These pass into the hilus of the lung with the bronchi and pulmonary vessels. Venous blood returns to the heart by way of bronchial veins which feed into the venae cavae or azygos veins.

Postganglionic sympathetic nerves supply the bronchi and stimulate a relaxation of smooth muscle (bronchodilation). Stimulation of **vagal fibers**, on the other hand, induces bronchoconstriction and increased secretion of mucous. Fine vagal filaments may also be found in the alveolar walls. These are related to stretch receptors, which sense the alveolar expansion during inspiration. In response to stimulation by these receptors, vagal fibers transmit the information to the respiratory center in the medulla, where inspiration is halted and expiration initiated. This center influences the firing of the phrenic and intercostal nerves and so maintains involuntary control over respiration.

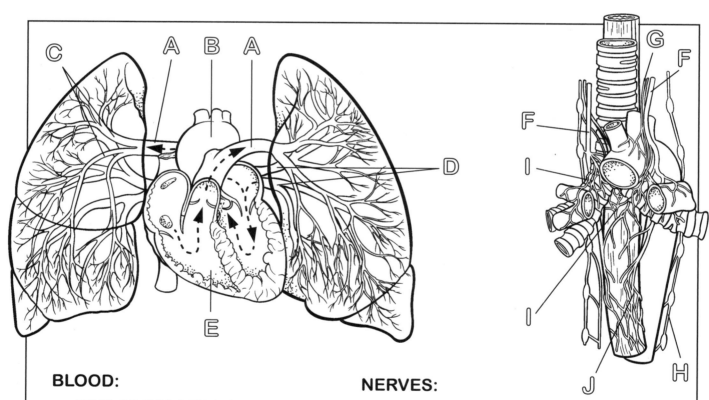

BLOOD:
A PULMONARY ARTERY
B AORTIC ARCH
C LOBAR ARTERIES
D PULMONARY VEINS
E HEART

NERVES:
F VAGUS
G CERVICAL CARDIAC
H SYMPATHETIC TRUNK
I PULMONARY PLEXUS
J ESOPHAGEAL PLEXUS

BRONCHI & THEIR FUNCTION

The main function of the bronchi is to provide passage for air to move in and out of each lung. Within each lung, multiple-branched bronchi may collectively be called a **bronchial tree**, consisting of the **primary bronchi**, which branch into the **secondary** and **tertiary bronchi**, and then branch into the **bronchioles**.

The lungs work with the circulatory system to pump oxygen-rich blood to all cells in the body. When you inhale, the bronchial tree delivers oxygen to tiny air sacs called **alveoli**, where **alveolus gas exchange** occurs: Inhaled oxygen diffuses from the alveoli into tiny **capillaries**, which is then carried from the lungs to the heart and thus to the rest of the body. Carbon dioxide (CO_2) is produced when cells use oxygen to get energy. The body rids itself of this byproduct via diffusion from the blood in the capillaries to the air in the alveoli, to the the bronchioles, bronchi, and back up and out as you exhale. There are around 1.5 million alveoli per lung!

*Note: Color the capillaries blue when **deoxygenated** and red when **oxygenated**.*

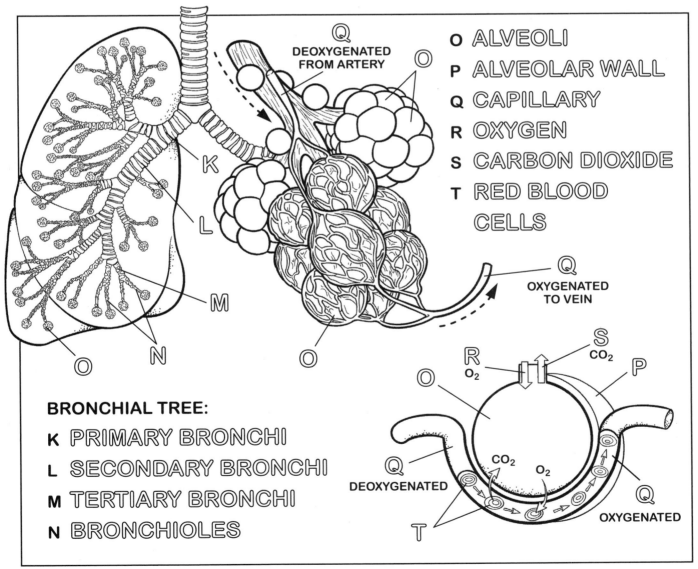

O ALVEOLI
P ALVEOLAR WALL
Q CAPILLARY
R OXYGEN
S CARBON DIOXIDE
T RED BLOOD CELLS

BRONCHIAL TREE:

K PRIMARY BRONCHI
L SECONDARY BRONCHI
M TERTIARY BRONCHI
N BRONCHIOLES

SECTION TEN:
INTERNAL ORGANS
OF THE ABDOMEN

This area is not simply "the stomach," as it is often referred to—though the stomach is indeed a resident. The abdomen consists of a great cavity filled with viscera and membranes bound by musculoskeletal walls.

Above, the abdomen is roofed by the thoracic diaphragm; below, it is continuous with the cavity of the pelvis. Anteriolaterally, the abdomen is bound by the lower extent of the thoracic cage, where musculotendinous sheets are interrupted in the midline by the straight muscle of the abdomen (the rectus abdominis). Posteriorly, the abdomen is supported by the stanchions of the lower thoracic and lumbar vertebrae and reinforced by the muscles of the posterior wall and back.

Interiorly, the cavity is lined with a layer of serous membrane (peritoneum) which is reflected onto and fitted tightly around most of the abdominal viscera—much as upholstery clothes a sofa. The viscera found in the abdomen consist of:

• A string of structures, the alimentary tract, continuing distally from the esophagus to the anus.
• Secretory organs associated with and embryologically derived from the alimentary tract.
• Certain endocrine organs.
• Structures associated with the urinary system.
• The spleen, part of the lymphatic system.

Included among these appropriate nerves, vessels, lymphatics, and ligaments, the contents of the abdomen are accounted for.

The organs of the abdomen are best described by the body system, as may be noted in the description of the digestive system in the previous section.

ABDOMINAL VISCERA OF THE DIGESTIVE SYSTEM
LIVER & GALLBLADDER

As discussed previously, the sections of the digestive system that reside in the abdominal viscera begin at the stomach and end at the anus. However, there are additional glands and organs that both directly and indirectly influence the digestive process. These include the liver and gallbladder, and on the next page, the pancreas and spleen.

LIVER

The **liver** (L., *hepar*) is the great gland of the abdomen. It weighs three pounds in a healthy individual—and can weigh more than ten pounds when diseased! It is reddish-brown as a result of extensive vascularity. It takes up much of the upper right and part of the upper left abdominal quadrant, tucked under the diaphragm and partly fenced in by the lower ribs. It consists of cords of epithelial cells attached to a tree of connective tissue, around which there is a network of **sinusoids** and capillaries.

The liver carries out many critical metabolic functions:
• Absorption of simple sugars; conversion to starch (glycogen) and storage of the same; breakdown of glycogen (glycogenesis) and release of simple sugar into the circulation.
• Absorption of protein and subsequent detachment and detoxification of nitrogen compounds, e.g., ammonia.
• Absorption and storage of fat; conversion of fat to glucose (gluconeogenesis).
• Detoxification of drugs; conversion of alcohol to carbohydrate (glycogen).
• Storage of vitamins A and D.
• Metabolism of products of red blood cell destruction; storage of iron.
• Production of a chemical soup called bile, which passes to the duodenum via the gallbladder to act on fats.
• Synthesis of urea, and material to be voided from the body by way of the kidneys.
• Synthesis of fibrinogen, a protein important in clotting.

All products of digestion pass out of the gastrointestinal (GI) tract into the liver for processing. They do so by way of a portal circuit of blood.

GALLBLADDER

The **gallbladder** is a reservoir of bile collected from the liver. The gallbladder communicates with the **common bile duct** via the **cystic duct**. The gallbladder receives diluted bile from the **hepatic ducts**, concentrates the bile by a 10:1 ratio, and stores it until a hormone from the **duodenum** (cholecystokinin) informs it to release the bile, which acts to break down the long-chain fat molecules so that absorption may occur.

Mineralization of bile may create "gallstones." Occlusion of the cystic duct and trauma to the mucosa trigger muscular spasms, and severe pain results.

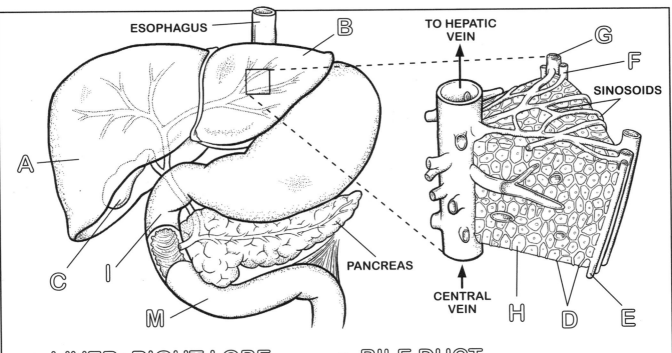

ESOPHAGUS

TO HEPATIC VEIN

SINOSOIDS

PANCREAS

CENTRAL VEIN

A LIVER: RIGHT LOBE
B LIVER: LEFT LOBE
C GALLBLADDER
D BILE CANALS

E BILE DUCT
F HEPATIC ARTERY
G PORTAL VEIN
H HEPATOCYTE

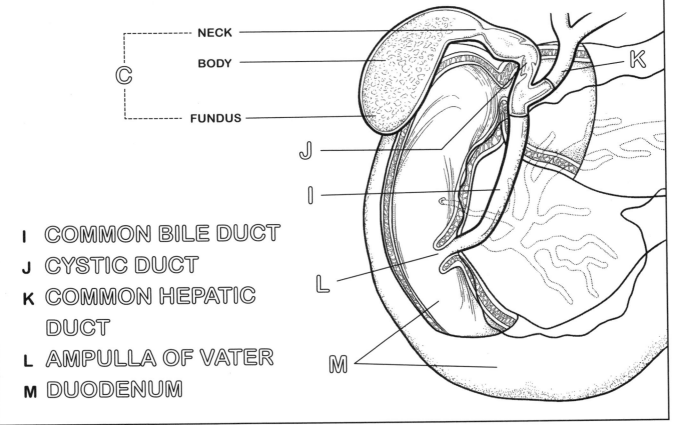

NECK

BODY

FUNDUS

I COMMON BILE DUCT
J CYSTIC DUCT
K COMMON HEPATIC DUCT
L AMPULLA OF VATER
M DUODENUM

ABDOMINAL VISCERA OF THE DIGESTIVE SYSTEM
PANCREAS & SPLEEN

PANCREAS

The **pancreas** is a soft, lobulated, overgrown salivary gland secured against the posterior abdominal wall by the peritoneum. The pancreas is basically two organs in one—an exocrine gland and an endocrine gland. The exocrine function (**acinar cells**) secretes enzymes (trypsin, amylase, and lipase) that break down proteins, carbohydrates, and fats. They are secreted in response to the hormones secretin and pancreozymin, which are released in the **duodenum** in response to foodstuffs entering from the stomach.

The endocrine gland portion of the pancreas consists of a population of **pancreatic islets** distributed apparently randomly throughout the gland. These cells produce and secrete two hormones, insulin and glucagon. Insulin, a polypeptide, influences the transport of sugar molecules desiring entry to the intracellular environment. Exhaustion of this chemical agent results in high concentrations of glucose in the blood (hyperglycemia) and spillage of glucose into the urine (glycosuria). Carbohydrate metabolism is altered abnormally and the resultant condition is termed diabetes mellitus. Glucagon, a polypeptide, acts on the liver cells to increase the breakdown of glycogen and so releases more sugar into the blood.

SPLEEN

The **spleen** is a very delicate, highly vascularized organ that is shaped like a baseball mitt without fingers. The spleen is an organ of the lymphatic system—in company with the thymus, tonsils, and lymph nodes. It is concerned with immunologic defense mechanisms and produces lymphocytes and monocytes, tools for both adaptive and innate immunity. The spleen acts as a filter of blood and a node that is a strainer for lymph.

The sinuses of the spleen are lined with phagocytic cells and here the blood is strained of old, degenerated blood cells as well as foreign matter. Filtered blood passes out of the splenic veins at the hilus.

Old red blood corpuscles are broken up and their iron stored by the splenic cells. Iron is transferred to the liver via the portal system in time of need. The pigment of discarded corpuscles is employed by the liver in the production of bile.

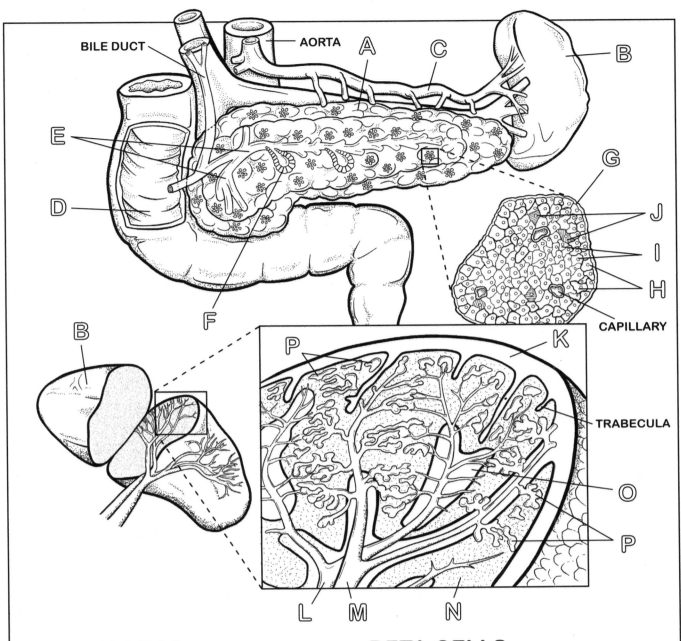

BILE DUCT

AORTA

CAPILLARY

TRABECULA

A PANCREAS
B SPLEEN
C SPLENIC ARTERY
D DUODENUM
E PANCREATIC DUCT
F ACINAR CELLS
G PANCREATIC ISLETS
H ALPHA CELLS

I BETA CELLS
J DELTA CELLS
K CAPSULE
L SPLENIC ARTERY
M SPLENIC VEIN
N RED PULP
O WHITE PULP
P VENOUS SINUSES

NERVE & VASCULAR SUPPLY OF THE ALIMENTARY TRACT & RELATED ORGANS

The nerve supply to the digestive organs is supplied and regulated by the autonomic nervous system, so digestion takes place without conscious effort, rather than willfully initiating each stage of the digestive process for survival.

The innervation plan of the alimentary canal may be seen on the adjacent illustration. Note that the parasympathetic nerve innervation of abdominal viscera is supplied by two different elements: The **vagus nerves** innervate the abdominal viscera to the level of the transverse colon, and the **pelvic splanchnic nerves** from the S2, 3, 4 levels of the spinal cord innervate all viscera below that level, including pelvic and perineal structures. The vagal fibers ramify within the aortic plexus and follow vessels out to the organs they supply. Within the walls of these organs, the pre- and postganglionic neurons and synapses are located. The pelvic splanchnic fibers reach the pelvic plexus and follow vessels to the organs they supply.

Sympathetic innervation of the abdominal viscera is derived from the **thoracic splanchnic nerves**. These pass through the diaphragm to enter the aortic plexus on the anterior aspect of the abdominal aorta. Within this plexus, usually associated with the principal visceral branches of the aorta, are sympathetic ganglia in which are found the preganglionic thoracic splanchnic nerves and synapses.

VASCULAR SUPPLY OF ABDOMINAL VISCERA

The arterial pattern of the abdominal viscera basically is supplied by three stems from the **abdominal aorta**, which supplies the viscera almost entirely. The **celiac artery** projects off the aorta just above the pancreas and immediately branches into the **hepatic, splenic**, and left **gastric arteries**. These branches supply the stomach, liver, gallbladder, duodenum, pancreas, and spleen. The **superior mesenteric** artery supplies the small intestine, appendix, cecum, and two thirds of the colon. The **inferior mesenteric artery** supplies the descending and sigmoid colon as well as the rectum.

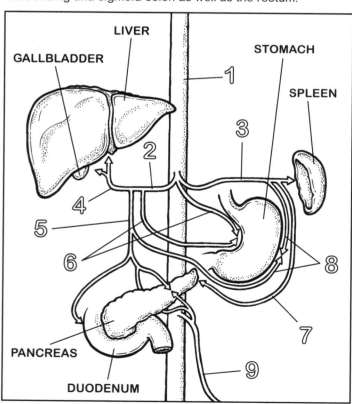

1 ABDOMINAL AORTA
2 COMMON HEPATIC ARTERY
3 SPLENIC ARTERY
4 HEPATIC ARTERY
5 GASTRODUODENAL ARTERY
6 GASTRIC ARTERIES
7 PANCREATIC BRANCH
8 GASTRO-OMENTAL ARTERY
9 SUPERIOR MESENTERIC ARTERY

A VAGUS NERVE
B PELVIC SPLANCHNIC
  NERVES
C THORACIC SPLANCHNIC
  NERVES
D LUMBAR/SACRAL
  SPLANCHNIC NERVES
E CELIAC GANGLION
  & PLEXUS
F AORTIC PLEXUS
G INFERIOR
  MESENTERIC
  ARTERY
H SUPERIOR
  MESENTERIC
  ARTERY
I RIGHT COMMON
  ILIAC ARTERY

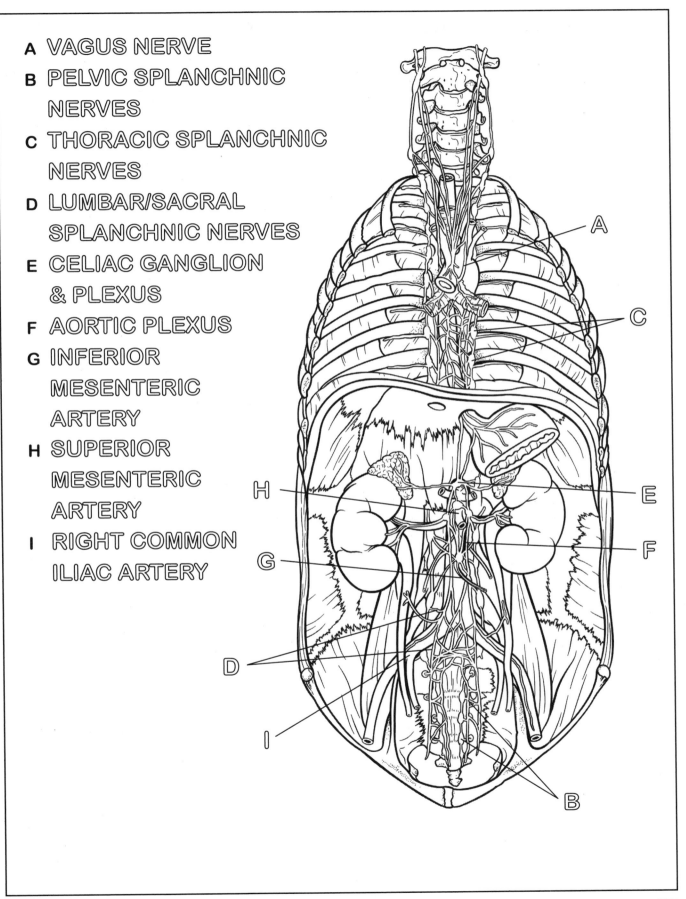

THE URINARY SYSTEM

The kidneys, ducts, and ureters reside in the abdominal region (viscera) while the bladder and urethra reside in the pelvic region (viscera), but for ease of understanding the **urinary system** is discussed here.

The urinary system consists of the paired **kidneys**, **ureters**, **urinary bladder**, and **urethra**. The structures function to maintain and adjust proper fluid and electrolyte balance through reabsorption and secretion as well as excretion of excess water and metabolites.

KIDNEYS

Like the liver in color and consistency, the brown, bean-shaped **kidneys** are fastened in a retroperitoneal position by a strong sheet of fascia reinforcing the peritoneum. The right kidney is lower than the left because of the imposing presence of the liver. The kidneys are divided into four major anatomical subdivisions:

• On the outer rind is the **cortex**. The cortex consists of tiny, convoluted tubules, some straight tubules, clusters of renal corpuscles, and a network of capillaries and arterioles.
• A middle belt of triangular **pyramids**—the **medulla**. Each pyramid may be seen to consist of longitudinal striations raining from the cortex. Columns of cortex dip down between medullary pyramids.
• An inner set of cup-like structures draining the pyramids—the **minor** and **major calyces**. The smaller calyces blend into larger calyces, which merge to form the pelvis and ureter beyond.
• The concavity of the kidney (hilus) disperses the **ureter** and **renal vein** and receives the **renal artery**.

ANATOMICAL BASIS OF KIDNEY FUNCTION

The functional unit of the kidney is the **nephron**. Demonstrating a veritable cornucopia of metabolic activity, the nephron is a filtering unit and consists of:

• A hollow epithelial ball (capsule) into which a tuft of arterioles (glomerulus) grow in embryonic development.
• A length of epithelial-lined tubing, (tubule) in both convoluted and straight segments, draining the capsule.
• A capsule and its glomerulus constitutes a renal corpuscle.

KIDNEY FUNCTION

Blood enters the kidneys through the renal artery where it is subjected to high-pressure filtration and passes into a long tubule. Here, the filtrate is treated to a most intense screening by the tubular epithelial cells. These cells can reabsorb elements of the filtrate and retain them, or they can transport them to blood capillaries or extracellular fluid adjacent to the tubules through the renal vein. Water and undesirable elements such as urea are reabsorbed by the tubular cells and removed from the body as urine. Sodium ions (Na+) are reabsorbed to provide electrical equilibrium.

URETER

The ureters are two muscular tubes with a pencil-point passageway from the kidneys to the bladder, and are lined with uroepithelium. They are eight to ten inches long and contain muscles that tighten and relax to force the urine down from the kidneys to the bladder.

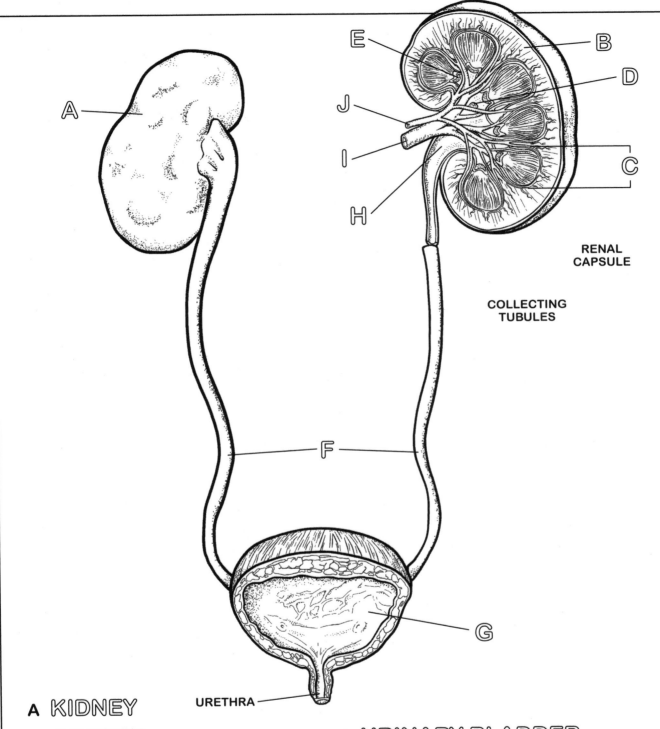

RENAL
CAPSULE

COLLECTING
TUBULES

URETHRA

A KIDNEY

B CORTEX

C MEDULLA

D MAJOR CALYCES

E MINOR CALYCES

F URETER

G URINARY BLADDER

H RENAL PELVIS

I RENAL VEIN

J RENAL ARTERY

ADRENAL GLANDS

The **adrenal** (suprarenal) **glands** sit atop the kidneys—enclosed in the same fascial investment—at a vertebral level of about T11. Triangular in shape, the adrenals are supplied by branches of local arteries including the aorta and the renal arteries, and drained by tributaries of the inferior vena cava.

When sectioned, the adrenals demonstrate an outer, thick, yellowish **cortex** and a thin, reddish **medulla**. The cortex consists of cords of epithelial cells arranged into three layers, and adjacent capillaries; the medulla, of irregular masses of epithelial cells in close contact with the capillaries.

The cortex secretes three groups of hormones, one of which is essential for life: glucocorticoids, mineralocorticoids, and sex hormones. Glucocorticoids, including cortisol and its metabolites, are related to carbohydrate metabolism, specifically, the breakdown of starch and the formation of glucose (gluconeogenesis). This activity is influenced by adrenocorticotropic hormone (ACTH) from the anterior lobe of the hypophysis. Mineralocorticoids, e.g., aldosterone, control water and electrolyte movement and balance by influencing tubular reabsorption and secretion. This activity is necessary for life. The cortex also secretes male sex hormones (testosterone) which may play a role in sexual development and which, if tumors develop, produce exaggerated masculinity in females. A function associated with the adrenal cortex that is not fully understood is resistance to stress. Following adrenalectomy, animals cannot withstand the rigors of cold or excessive heat; they cannot cope with "flight-or-fight" situations and in response to these crises, they will die.

The **medulla** secretes epinephrine and norepinephrine. These hormones activate the sympathetic "flight-or-fight" response as well as affect carbohydrate metabolism. Norepinephrine is also secreted by sympathetic preganglionic neurons at synapses.

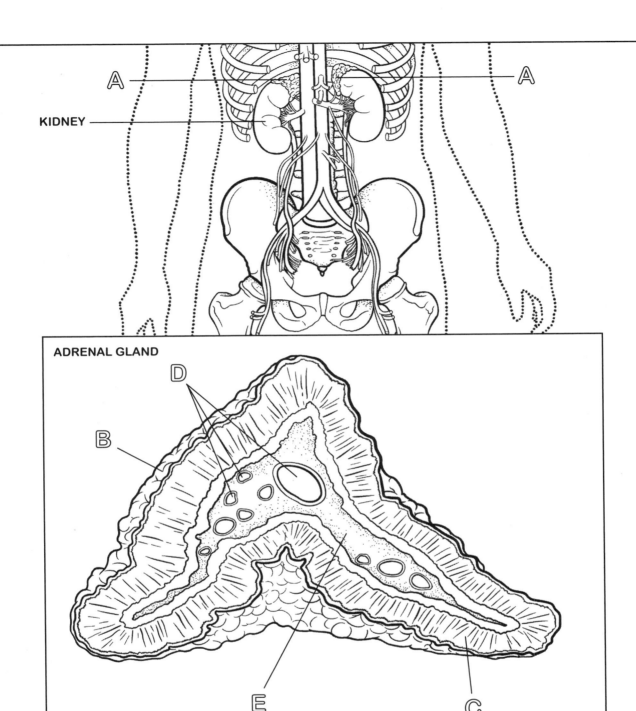

KIDNEY

ADRENAL GLAND

A ADRENAL GLAND
B CAPSULE
C CORTEX
D BLOOD VESSELS
E MEDULLA

SECTION ELEVEN:
THE PELVIS

The pelvis is a cavity bounded anteriorly and laterally by the paired hip bones, and posteriorly by the sacrum and coccyx. The pelvis is a direct inferior extension of the abdominal cavity and, therefore, has no roof.
It contains:

- Internal reproductive structures.
- The urinary bladder.
- The rectum.
- Assorted ducts, vessels, and nerves.

PELVIC BOUNDARIES

Pelvic viscera—the **bladder**, **rectum**, pelvic **genital organs,** and terminal of the **urethra**—all reside within the **pelvic cavity** (or the true pelvis). The **pelvic diaphragm** is formed by the **coccygeus** and **levator ani** muscles, separating the pelvis above from the **perineum** below.

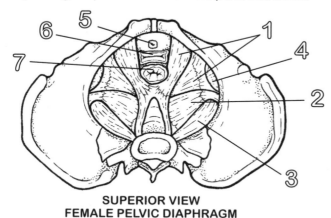

SUPERIOR VIEW
FEMALE PELVIC DIAPHRAGM

1 LEVATOR ANI
2 COCCYGEUS
3 PIRIFORMIS
4 OBTURATOR
5 URETHRA
6 VAGINA
7 ANAL CANAL

The perineum is the region within the framework of the bony pelvic outlet. Its floor of skin and fascia—perforated by structures descending from the pelvis—is continuous with the skin and fascia of the anterior abdominal wall, the buttocks, and the thighs.

The perineal spaces are largely fat-filled and encompass:
• The urethra, passing distally from the bladder to the exterior.
• The anal canal and anus, passing distally from the rectum to the exterior.
• In the female, the distal continuation of the vagina.
• Miscellaneous glands, muscles, ducts, vessels, and nerves.

The pelvis is divided into a greater space above, the pelvis major, and a smaller space below, the pelvis minor, by a line circling posteriorly from the upper surface of the pubic bone to the sacrum. This is the pelvic brim.

The lateral walls of the lesser pelvis minor consist of:
• The interior surface of the superior pubic ramus.
• Part of the iliac bone.
• Part of the **obturator's internus muscle** covering up to the obturator canal.
• Part of the **levator ani**.
• Spaces of the sciatic foramina filled with structures exiting and entering laterally.

The hammock-like **levator ani** rises from the ilium and **pubic bone** and extends posteromedially to insert on the coccyx via the anococcygeal ligament, forming a sling around the rectum. The levator ani and its fascial envelope constitute the pelvic diaphragm.

The **rectum** (L., *rectus*, straight) is the extension of the **large intestine**, following the more S-shaped (**sigmoid**) **colon** of the abdomen. The rectum begins roughly at the midsacral level and, following the curve of the sacrum, becomes the anal canal as it pierces the levator ani.

The anal canal represents the last inch and a half of some twenty feet of digestive tubing. The most distal part of the large intestine, it is surrounded by the external sphincter within the ischiorectal fossa of the perineum and opens to the exterior as the **anus**.

A OVARY
B UTERUS
C RECTUM
D ANUS
E VAGINA
F URETHRA
G PUBIC BONE
H URINARY
   BLADDER
I URETER
J SEMINAL
   VESICLES
K DUCTUS
   DEFERENS
L PROSTATE
M SIGMOID COLON
N LEVATOR ANI
   MUSCLE
O OBTURATOR
   INTERNUS
   MUSCLE

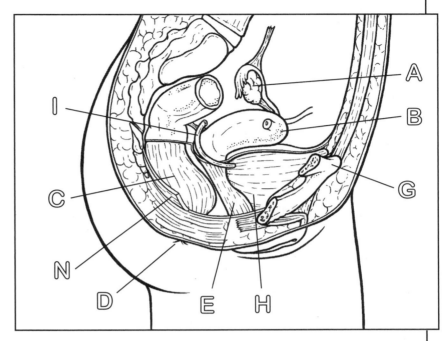

FEMALE

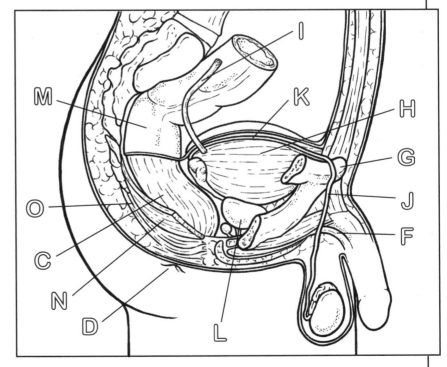

MALE

VESSELS & NERVES OF THE PELVIS & PERINEUM

The principal source of blood to the pelvis and perineum are the **internal iliac arteries**, one of two terminal branches of the **common iliac arteries**. The internal iliac artery—on each side—curves medially and downward after leaving the common iliac, to enter the true pelvis. It divides irregularly into visceral and somatic viral branches. The visceral branches include:

• The **umbilical artery**, which in the fetus passes up and through the umbilical cord to supply the placenta with unoxygenated blood). It is partly obliterated back from the umbilicus to the bladder and sends off branches to that organ (superior vesical arteries).
• The **inferior vesical artery**, which serves the bladder, seminal vesicles, prostate, and ductus deferens.
• The **middle rectal artery** to the rectum—one of three arteries to that organ.
• In the female, the homologue of the male's inferior vesical arteries become the **uterine** and **vaginal arteries**.

Blood to the perineum arrives through the **internal pudendal artery**, another branch of the internal iliac. The external iliac artery contributes some branches to the scrotum and labia. The veins, in general, follow the arteries and form dense plexuses within the pelvis. The most significant plexus from a clinical standpoint is that associated with the **rectal veins**. Tributaries of these veins in the anal columns are often subject to trauma by hard fecal stools contributing to inflammation of these veins. As they heal, scar tissue develops. With recurring inflammation the veins become twisted and tortuous (hemorrhoids).

NERVES

Visceral structures are supplied by autonomic nerves, musculoskeletal structures by somatic (spinal) nerves. Pelvic viscera receive their motor and sensory innervation via the pelvic plexus. This autonomic plexus receives parasympathetic fibers from S2-4 levels of the spinal cord via **pelvic splanchnic nerves**. It receives sympathetic fibers from the aortic plexus via the hypogastric plexus. The pelvic splanchnic nerves bring about contraction of the bladder muscles and dilation of the arteries to the penis/clitoris. Since this latter action causes erection, these nerves are historically referred to as "*nervi erigentes*." The sympathetic nerves function to initiate emission and pre-ejaculation by causing contraction of the muscular walls of the **ductus deferens** and the prostatic/seminal gland musculature. Sympathetic nerves also function to constrict arteries in the pelvic and perineal areas.

The **pudendal nerve** from spinal segments S2-4 supplies the skeletal muscles of the perineum.

LYMPHATIC DRAINAGE

Lymph nodes are generally arranged around the major arteries of the pelvis. The principal collection of nodes within the pelvis is located about the **sacral artery** and the common, external, and internal iliac arteries. These drain the pelvis and some perineal structures. The deep and superficial **inguinal nodes** drain most of the perineum. The pelvic and inguinal nodes drain toward the common iliac nodes, which drain into the lumbar trunks. These trunks flow into the **cisterna chyli**, which is the distal extremity of the thoracic duct.

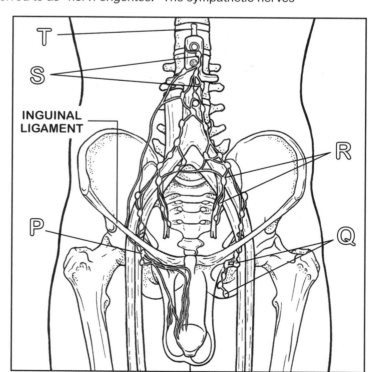

INGUINAL LIGAMENT

ARTERIES (VEINS ARE BEHIND):

A COMMON ILIAC
B INTERNAL ILIAC
C EXTERNAL ILIAC
D SUPERIOR GLUTEAL
E UMBILICAL
F OBTURATOR
G INFERIOR VESICULAR
H UTERINE
I MIDDLE RECTAL
J INTERNAL PUDENDAL

NERVES:

K PELVIC SPLANCHNIC
L PUDENDAL
M TERMINAL BRANCHES OF PUDENDAL NERVE
N PELVIC PLEXUS
O HYPOGASTRIC NERVES

LYMPH NODES (OPPOSITE PAGE):

P DEEP INGUINAL
Q SUPERFICIAL INGUINAL
R INTERNAL ILIAC
S PRE-AORTIC
T CISTERNA CHYLI

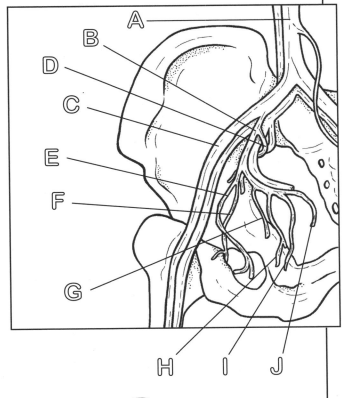

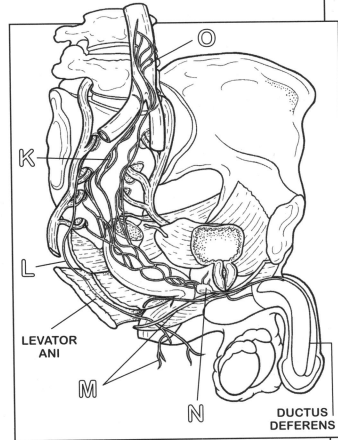

LEVATOR ANI

DUCTUS DEFERENS

MALE ORGANS OF REPRODUCTION: TESTES

The principal function of the male reproductive system is to produce germ cells, or **sperm**. Thus the primary sex organ of the male is the pair of structures that produce sperm cells: the **testes** (singular, **testis**). The secondary but equally important function is to conduct those sperm cells from the site of development (the testes) to the outside of the body. Thus, all the glands and ducts associated with this movement are called **accessory structures** of reproduction.

Initially developed within the abdominal cavity at the level of the kidneys, each testis descends through the anterior abdominal wall via the inguinal ring and canal on its side and enters the **scrotum**. The canal is occupied by the **ductus deferens**, the coverings it took along with it during descent, and related **vessels** and **nerves**. These structures collectively make up the **spermatic cord**.

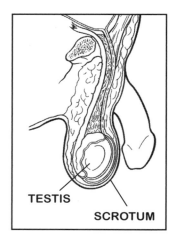

TESTIS

SCROTUM

TESTES
The testes are walnut-shaped structures situated in the scrotum. There are normally two of them and the left usually hangs lower than the right. Each testis has a superior and inferior pole (end) and the **epididymis** sits atop the former. The outer covering of the testis is a thick fibrous capsule—the **tunica albuginea**—from which septa (walls) of connective tissue project into the lumen of the testis and compartmentalize it. A number of highly coiled **seminiferous tubules** are tucked away into each compartment.

The primordial germ cells are the **spermatogonia** at the periphery of the tubules and develop into **primary spermatocytes**. At this point, each of these cells has a full complement of chromosomes (forty-six). The primary spermatocytes undergo a type of division into **secondary spermatocytes**, each of which has half of the usual chromosome count (twenty-three). This process is termed *meiosis*, and it is a critical stage in sperm development. The secondary spermatocytes divide (mitotically) into **spermatids** which mature into **spermatozoa**. These are now mature sperm, though generally inactive, and find their way to the epididymis to await a call to active duty.

INTERSTITIAL CELLS
Interspersed about the sparse connective tissue among the **seminiferous tubules** is a significant collection of specialized cells and associated capillaries. These **interstitial cells** are known to secrete the male sex hormone **testosterone**, which not only stimulates the appearance of secondary sex characteristics at puberty but also has a significant effect on growth in the adolescent. Interestingly, testosterone does not have the role of maintaining the sperm-growing epithelium. What does? FSH.

What are some of the secondary sex characteristics brought on by testosterone?

• Increase in size of the penis and development of the testes.
• Maturation and increased activity of the reproductive glands, i.e., prostate, seminal vesicles.
• An increase in muscle size and strength.
• Initiation of sex drive (libido).
• Deeper voice due to changes in laryngeal structure.
• Increase in and thickening of sebaceous gland secretions.
• Increase in body hair, especially in pubic and axillary regions beginning at puberty.
• Later in life, a possible effect on balding.

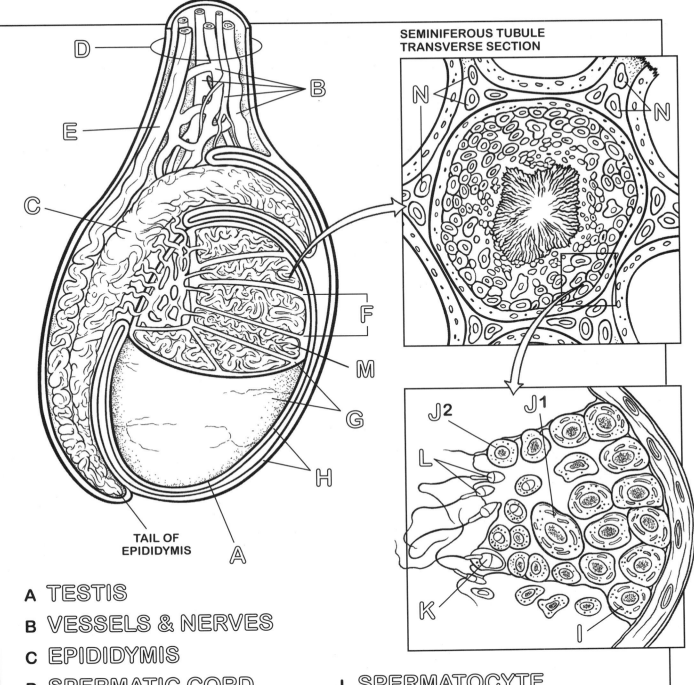

SEMINIFEROUS TUBULE
TRANSVERSE SECTION

TAIL OF
EPIDIDYMIS

A TESTIS
B VESSELS & NERVES
C EPIDIDYMIS
D SPERMATIC CORD
E DUCTUS DEFERENS
F SEMINAL VESICLE
   LOBULES
G TUNICA ALBUGINEA
H TUNICA VAGINALIS
I SPERMATOGONIUM

J SPERMATOCYTE
   J1 PRIMARY
   J2 SECONDARY
K SPERMATIDS
L SPERMATOZOA
M SEMINIFEROUS TUBULES
N INTERSTITIAL CELLS

MALE ORGANS OF REPRODUCTION: ACCESSORY STRUCTURES

EPIDIDYMIS

As we have seen, the **epididymis** sits atop each testis, consisting of a highly convoluted tubule in which inactive living sperm are stored. The tubule descends along the posterior aspect of the testis to the level of the inferior pole, where it turns upward to become the **ductus deferens**.

Microscopically, the tubule is lined with pseudostratified epithelium whose cilia are nonmotile. These cilia secrete glycogen droplets for the substance of stored spermatozoa. During sexual activity, the sperm are moved out of the epididymis by contractions of smooth muscle fibers oriented about the tubule.

DUCTUS DEFERENS AND EJACULATING DUCTS

The ductus deferens is the principal and largest duct of the male reproductive system. It arises at the tail of the epididymis, passes out of the **scrotum** to enter the anterior abdominal wall at the superficial inguinal ring, transverses the inguinal canal in company with related vessels and nerves as the spermatic cord, and enters the abdominal cavity (via the deep inguinal ring) to arch across the side of the pelvis and pass behind the bladder. Finally, it perforates the **prostate gland**, becomes narrower and receives the duct of the **seminal vesicle**, and joins the prostatic **urethra** as the pencil-point in the **ejaculatory duct**.

PROSTATE AND SEMINAL VESICLES

Collectively these structures are classified as **accessory genital glands**. The seminal vesicles are actually appendages of the ductus deferens, consisting of an elongated sac lined with secretory epithelium. The prostate, on the other hand, is a mass of tubuloalveolar glands (within a connective tissue capsule) whose ducts open into the urethra. In both glands, smooth muscle contributes heavily to their structure.

The seminal vesicles, located behind the bladder, can often be palpated through the rectum. The prostate is at the base of the bladder and encloses the upper (prostatic) urethra. Both glands actively secrete milky, alkaline material rich in fructose (sugar) which aids sperm motility.

The seminal vesicle empties its secretions into the ductus deferens. The prostate receives the urethra (from the bladder) as well as the ejaculatory ducts and empties its secretions directly into the urethra. Descending through and exiting anteroinferiorly from the prostate, the urethra penetrates the levator ani to enter the perineum.

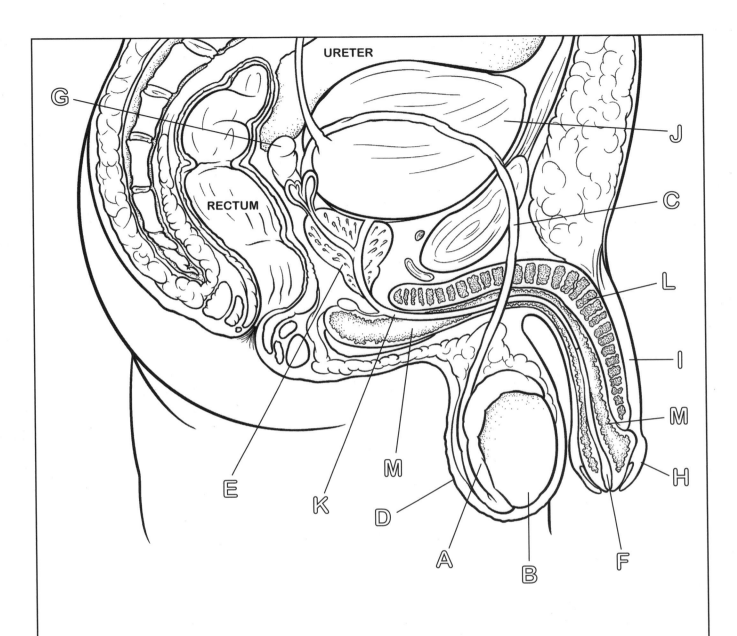

URETER

RECTUM

A EPIDIDYMIS	**G** SEMINAL VESICILE
B TESTIS/TESTICLE	**H** PREPUCE (FORESKIN)
C DUCTUS (VAS) DEFERENS	**I** PENIS
D SCROTUM	**J** URINARY BLADDER
E PROSTATE GLAND	**K** EJACULATORY DUCT
F URETHRA	**L** CORPUS CAVERNOSUM
	M CORPUS SPONGIOSUM

FEMALE ORGANS OF REPRODUCTION: OVARIAN FUNCTION

Once again we can look at the reproductive system from a conceptual point of view and find that the functional and anatomical order of structures in the female is entirely reasonable. To conceive a human being, a male germ cell must combine with a female germ cell. We have seen in the male a primary sex organ (testis) that generates sperm (germ) cells, and a set of ducts and glands providing the wherewithal for those sperm to be transported out of the body.

The meeting place for the successful union of the sperm with a female germ cell oval is critical. The place must be warm and moist, and a host of hormones and nutrient media must be on hand. Therefore, an organ carrying on this activity must be (1) located within the body, (2) connected to the outside via a duct so that the sperm can reach it, and (3) connected to the organ of female cell generation by a duct.

The adjacent illustration demonstrates that the actual anatomy indeed meets all of these criteria, and more: The **uterus**, the site of embryonic and fetal development, is located within the pelvic cavity. It is an internal structure (warm), glandular (moist), and richly supplied with blood (carrying hormones).
The uterus is in communication with the outside by way of a duct, the **vagina**, which receives the sperm of the male. The uterus is in communication with the primary sex organs, the **ovaries**, via bilateral **uterine tubes**.

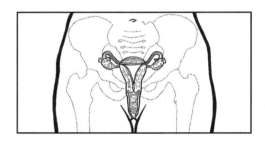

OVARY

The **ovaries** (paired) are located on the lateral walls of the pelvis at the level of the anterior superior iliac spine. These structures arise with the same tissue in the same place as the male testes, hence ovaries and the testes look alike externally, although the former is smaller, flatter, and in the adult, scarred from numerous previous ovulations.

OVARIAN FUNCTION

The ovary really has a two-fold function: to produce female germ cells (**ovum** or eggs) and produce two hormones which act to develop the uterine lining, the **endometrium**, in preparation for implantation of the embryo. The ovary performs these functions cyclically under the direction of hormones from the anterior lobe of the pituitary gland. It takes about twenty-one days for the ovary to develop a primordial **follicle** into a mature one with an ovum, ready to blast off for the uterine tubes. In the next ten days, the ovary is concerned with secreting hormones which maintain the uterine lining at a nutritious level for the fertilized ovum. The cycle of the ovary does not correspond precisely to that of the uterus (menstrual cycle), and that is understandable if one remembers that the development of the uterine lining is dependent upon ovarian secretions which necessarily precede it.

The ejection of the ovum is termed **ovulation**, and is believed to be a consequence of the combined action of FSH, estrogen, and another pituitary product, luteinizing hormone.

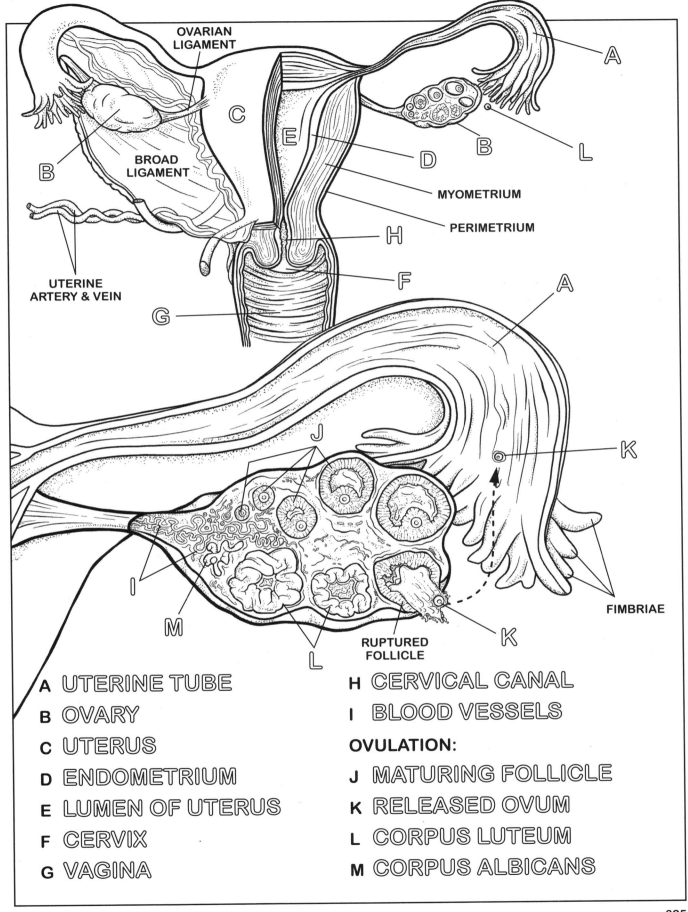

OVARIAN
LIGAMENT

BROAD
LIGAMENT

UTERINE
ARTERY & VEIN

MYOMETRIUM

PERIMETRIUM

FIMBRIAE

RUPTURED
FOLLICLE

A UTERINE TUBE

B OVARY

C UTERUS

D ENDOMETRIUM

E LUMEN OF UTERUS

F CERVIX

G VAGINA

H CERVICAL CANAL

I BLOOD VESSELS

OVULATION:

J MATURING FOLLICLE

K RELEASED OVUM

L CORPUS LUTEUM

M CORPUS ALBICANS

FEMALE ORGANS OF REPRODUCTION: THE UTERUS & VAGINA

UTERINE TUBES

The uterine tubes are projections of the uterus and conduct the ejected ova from the ovary to the uterus. The tubes are draped with peritoneum which extends from the uterus to the lateral pelvic wall.

The mucous membrane of the tubes is characterized by a folded and irregular ciliated columnar epithelium supported by a highly vascular connective tissue, suggesting secretory activity. And indeed, the epithelial cells are secretory, creating a nutrient environment for the ovum that descends toward the uterus. The cilia beat toward the uterine lumen, aiding the ovum in its migration. The mucous membrane is reinforced by a layer of smooth muscle, which probably also aids the ovum's movement.

Sperm cells, when deposited in the vagina, rapidly (within twelve to twenty-four hours) migrate to the uterine tubes and there can fertilize an egg. A fertilized ovum will then quickly divide into two, four, eight, and then many cells, moving down the fallopian tube and implanting in the uterus—where an embryo can start growing.

UTERUS

The **uterus**, also referred to as the womb, provides a secure residence in which the developing fetus completes the nine-month gestational period. Embryologically, the uterus represents the fusion of the two uterine tubes. It is a pear-shaped, muscular organ, no bigger than a closed fist.

The uterus has three principal parts: the **fundus** above, the main **body** and lower **isthmus**, and the lower, narrower **cervix**.

The uterus is continuous with the uterine tubes at the fundus and in communication with the **vagina** at the cervix. The neighbors of the uterus include the rectum posteriorly (with some ileum intervening) and the bladder immediately anterior. Laterally, the uterus is related to the broad ligaments and the structures therein. Ligaments within the broad ligament are responsible for stabilizing the uterus, although the main support probably comes from its attachment to the vagina.

VAGINA

The vagina is the access to the uterus and its tubes from the external environment. A collapsed fibromuscular passage, it embraces the penis in sexual intercourse, recovers the semen following ejaculation, and is capable of great distention for the newborn, who must pass through it to enter "the real world."

Superiorly, the vagina receives the cervix of the uterus; inferiorly, the vagina is open to the outside within the vestibule of the perineum. Here the vagina may be partially covered by the hymen, a mucous membrane fold that can stretch or tear during first penetrative intercourse (though this is not the only way the hymen can stretch or tear). Some are born with little or (rarely) no hymenal tissue.

The vagina is supplied by branches of the **uterine artery** (upper part) and by direct branches of the **internal iliac artery**. The upper part of the vagina is served by an autonomic plexus, while the lower part is serviced by the **pudendal nerve**.

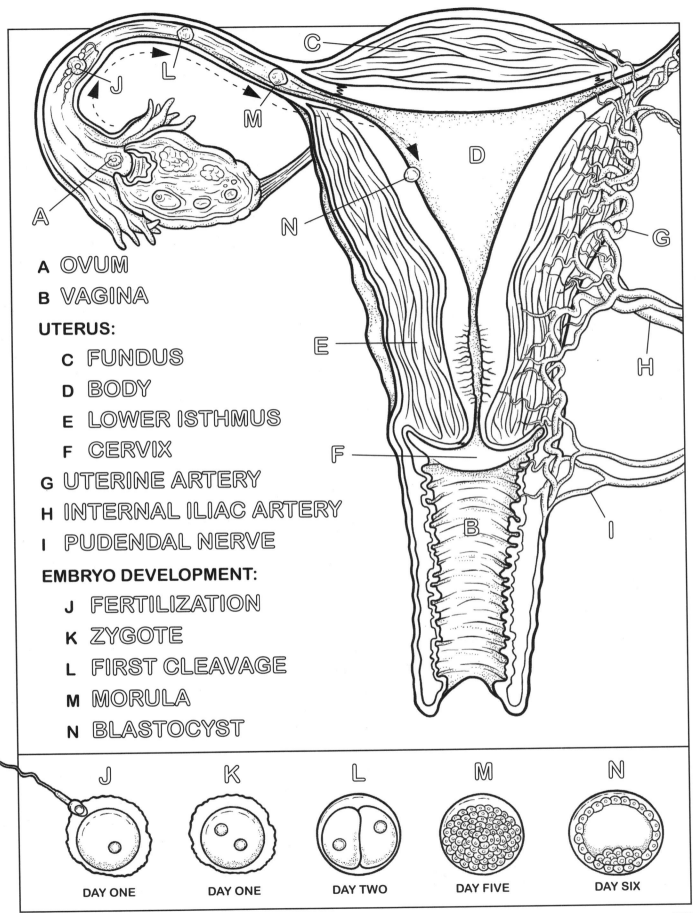

A OVUM
B VAGINA
UTERUS:
C FUNDUS
D BODY
E LOWER ISTHMUS
F CERVIX
G UTERINE ARTERY
H INTERNAL ILIAC ARTERY
I PUDENDAL NERVE
EMBRYO DEVELOPMENT:
J FERTILIZATION
K ZYGOTE
L FIRST CLEAVAGE
M MORULA
N BLASTOCYST

J DAY ONE
K DAY ONE
L DAY TWO
M DAY FIVE
N DAY SIX

PERINEUM

A CONCEPTUAL ANALYSIS

Anatomically, the **perineum** region is complicated but the general organization can be learned without difficulty.

The perineum may be visualized as a box rotated 45° to form a three-dimensional, diamond-shaped structure. This is the basic shape of the perineum within the framework of the pelvic outlet. Thus the four corners of the perineum are anteriorly, the **pubic symphysis**; posteriorly, the **coccyx**; and laterally, the **ischial tuberosities**. The four sides of the perineum may be seen to be the **ischiopubic bones** anterolaterally, and the **sacrotuberous ligaments** posterolaterally. The roof of the perineum is the **levator ani** or pelvic diaphragm. The floor of the perineum is created from skin and superficial fascia and is perforated by passageways from the pelvis.

The four-sided perineum may be bisected into two triangles, anterior and posterior. The anterior triangle contains the ducts of the urinary and genital systems and is, therefore, conveniently titled the **urogenital triangle**. The posterior triangle encompasses the anal canal and is so called the **anal triangle**.

UROGENITAL TRIANGLE

The urogenital (UG) region of the perineum is a space like the ischiorectal fossa with a pertinent exception: A shelf exists that separates the UG region into upper and lower compartments. Such a shelf is properly called a diaphragm, specifically the **urogenital diaphragm**. The upper compartment is continuous with the more posterior ischiorectal fossa; it is, therefore, referred to as the **anterior recess of the fossa**. The recess is split into two side-by-side recesses by the midline perineal body and the low-slung levator ani, which supports the urethra, the prostate (in the male), and the vagina.

The lower compartment is the superficial perineal space, containing muscular and specialized structures related to the external genital organs.

The UG diaphragm itself, therefore, must be, and is, the deep perineal space.

ANAL TRIANGLE

The space of this triangle is called the **ischiorectal fossa**, and its principal occupant is the **anus**. The rectal fossa tissue is largely fat filled. Under pressure of sitting, because it is relatively poorly vascularized, the fossa is an occasional site of abscess. On the lateral walls of the fossa are the fascia-formed pudendal canals connecting nerves and vessels to the perineum from the pelvis via the deep gluteal region.

At the anterior border of the triangle, arising as deep as the levator ani, there is a fibromuscular mass, the perineal body. It functions as a site of attachment and support for many of the muscles of the perineum. If it should be damaged during childbirth, the stability of the passageways of the perineum may be in danger.

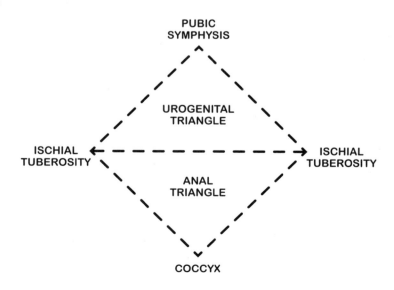

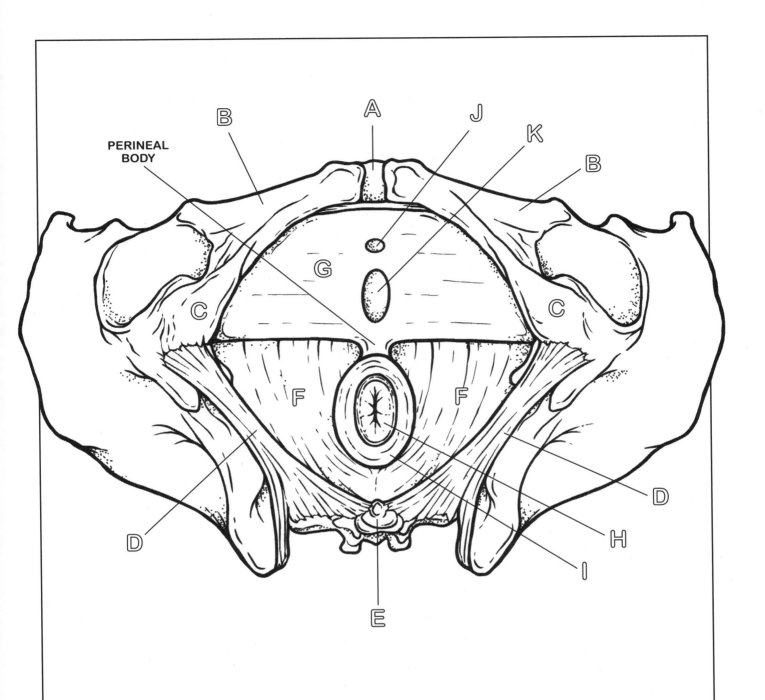

PERINEAL
BODY

A PUBIC SYMPHYSIS
B ISCHIOPUBIC BONE
C ISCHIAL TUBEROSITY
D SACROTUBEROUS
   LIGAMENT
E COCCYX

F LEVATOR ANI
G UROGENITAL DIAPHRAGM
H ANUS
I ANAL SPHINCTER MUSCLE
J URETHRAL ORIFICE
K VAGINAL ORIFICE

THE MALE UROGENITAL REGION

The superficial perineal space of the male includes:
• Roots of the penis.
• Muscular envelopes of these roots.
• Related nerves and vessels.

The deep perineal space consists entirely of the urogenital diaphragm. The penis has three fibroelastic roots. Two of these, the **crura**, arise from the underside of the ischial bones. The other, the **bulb**, arises between the crura at the urogenital diaphragm. These erectile bodies, sheathed in skeletal muscle, are properly part of the superficial perineal space. The bulb of the penis incorporates the **urethra** and is wrapped in the **bulbospongiosus** muscle, which has its origin in the perineal body. This muscle functions in erection and ejaculation and receives innervation from the pudendal nerve.

Each crus of the penis is wrapped in the **ischiocavernosus** muscle, arising from the ischium. These muscles, which aid in erection, receive innervation from the pudendal nerve.

MALE EXTERNAL GENITAL ORGANS
The external genital organs are the **penis** and **scrotum**.

PENIS
The **penis** permits introduction of sperm into the vagina. Since the vagina is collapsed, the penis must be erect in order to penetrate—and this is what the morphology of the penis is all about. The bulb of the penis arises between the two crura from the base of the urogenital diaphragm. The two crura and bulb come together to form the triple-barreled body of the penis. Where the body bends downward the bulb is renamed the **corpus spongiosum**, and the two crura become the **corpora cavernosa**. The corpus spongiosum terminates distally by flaring out as the **glans**. The two-layered skin variably covering the glans is called the prepuce and is sometimes removed at birth (circumcision).

SCROTUM
The **scrotum** is a thin-skinned pouch for the testes. It is borrowed from the anterior abdominal wall and is characterized by:
• Dark pigment.
• A tunic of smooth muscle, the **dartos muscle**, associated with the nonfatty superficial fascia.
• A median septum dividing the scrotum into two compartments. A seam, or **raphe**, visible in the midline of the scrotum, can be seen to continue up the underside of the skin of the penis to the glans.

The scrotal skin has the capacity to change its appearance and physical state in response to the action of the dartos muscle, the aim being to maintain the proper temperature within the scrotum. In cold, the muscle contracts and the scrotum wrinkles, drawing more closely around the testes and so stemming the loss of heat. In warmth, the dartos relaxes, the scrotum loses its tone, and heat is allowed to dissipate.

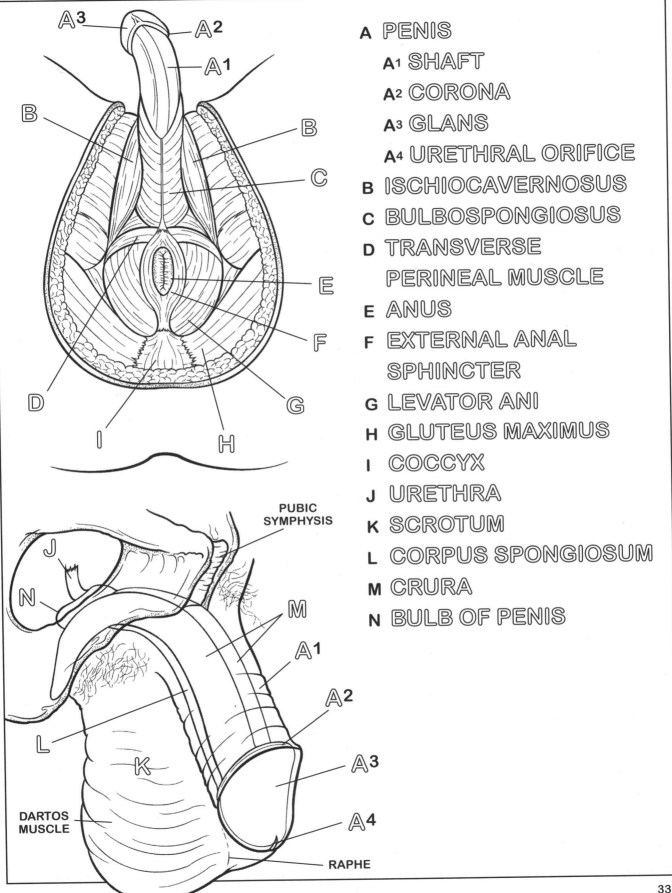

A PENIS
   A₁ SHAFT
   A₂ CORONA
   A₃ GLANS
   A₄ URETHRAL ORIFICE
B ISCHIOCAVERNOSUS
C BULBOSPONGIOSUS
D TRANSVERSE
   PERINEAL MUSCLE
E ANUS
F EXTERNAL ANAL
   SPHINCTER
G LEVATOR ANI
H GLUTEUS MAXIMUS
I COCCYX
J URETHRA
K SCROTUM
L CORPUS SPONGIOSUM
M CRURA
N BULB OF PENIS

PUBIC
SYMPHYSIS

DARTOS
MUSCLE

RAPHE

331

THE FEMALE UROGENITAL REGION

The anatomy of the superficial and the perineal spaces here is different from the male by the presence of the vagina and associated external genital organs. Otherwise, the anatomy is similar if not homologous. The urogenital diaphragm is pierced by the urethra anteriorly and the vagina posteriorly. The contents of the urogenital diaphragm are similar to those of the male, including deep transverse perineal muscles which help stabilize the important perineal body, and the inconsequential "sphincter of the urethra."

There is one addition—the **vagina**—and one subtraction—the bulbourethral glands.

The muscles of the superficial perineal space are homologous with those of the male; except that the **bulbospongiosus muscle** (overlying the **vestibular bulb**) is cleaved as it circumvents the vagina. The superficial transverse muscles appear to stabilize the **perineal body**. If these muscles were cut in an episiotomy procedure, the stability of the perineal body might be in question.

The **pudendal nerve** and vessels to the perineum arrive from the pelvis via the pudendal canal of the ischiorectal fossa.

FEMALE EXTERNAL GENITAL ORGANS

The external genital organs (vulva) are the:
• **Labia majora** or large lips.
• **Labia minora** or small lips.
• **Clitoris**.
• **Vestibule** or space created by the small lips.
• **Bulbs** of the vestibule.
• Greater **vestibular glands** (next to the **vaginal orifice**).

The **labia majora** consists of two fat-filled folds of skin, covered with hair (beginning at puberty). The **labia minora** are folds of skin enclosing the **vaginal opening**. The **clitoris**—homologous of the penis—consists of two erection bodies (corpora cavernosa clitoridis), which arise from the ischial rami. Like the penis, the clitoris is very sensitive to touch and pressure and is capable of erection. The bulbs of the vestibule consist of a pair of erectile bodies on either side of the vagina. They are homologous with the bulb of the penis. The bulbs are capable of significant distention during sexual stimulation, enhancing vaginal contact with the penis. The vestibular glands are pea-sized structures adjacent to the bulbs. They are homologues of the bulbourethral glands, and empty their secretions at the margin of the vaginal orifice.

VULVA:

1 CLITORIS

2 HOOD OF CLITORIS

3 LABIA MAJORA

4 LABIA MINORA

5 VAGINAL OPENING

6 URETHRA

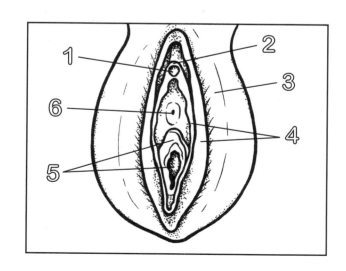

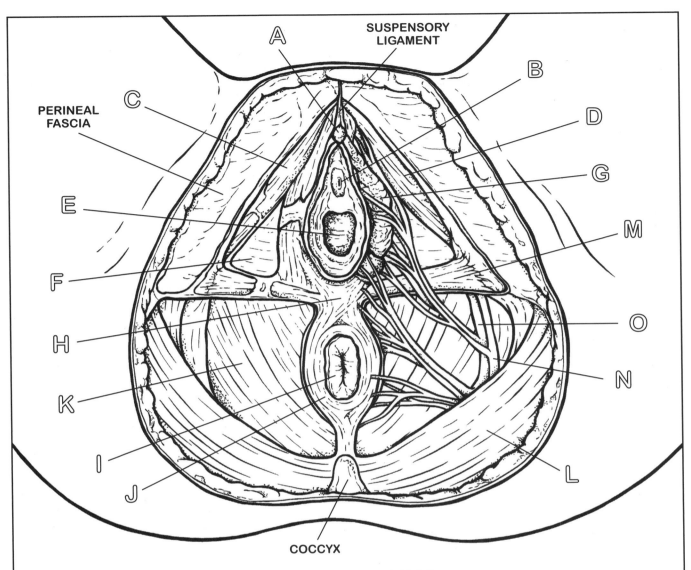

SUSPENSORY LIGAMENT

PERINEAL FASCIA

COCCYX

A CLITORIS
B URETHRA
C BULBOSPONGIOSUS MUSCLE
D ISCHIOCAVERNOSUS MUSCLE
E VAGINAL OPENING
F PERINEAL MEMBRANE
G VESTIBULAR BULB
H PERINEAL BODY

I ANUS
J ANAL SPHINCTER
K LEVATOR ANI MUSCLE
L GLUTEUS MAXIMUS
M TRANSVERSE PERINEAL MUSCLE
N PUDENDAL NERVE & BRANCHES
O PUDENDAL ARTERY & BRANCHES

INDEX

INDEX

INDEX

INDEX

INDEX

INDEX

ENJOY & LEARN WITH OTHER BOOKS
FROM COLORING CONCEPTS

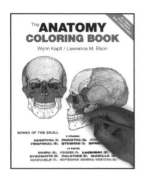

**The Anatomy
Coloring Book**

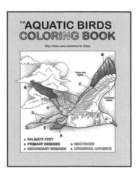

**The Aquatic Birds
Coloring Book**

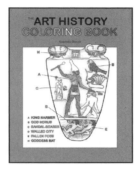

**The Art History
Coloring Book**

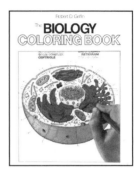

**The Biology
Coloring Book**

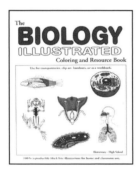

**The Biology
Illustrated Coloring
and Resource Book**

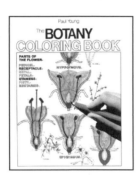

**The Botany
Coloring Book**

**The Human Brain
Coloring Book**

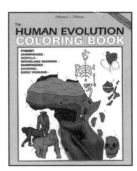

**The Human Evolution
Coloring Book**

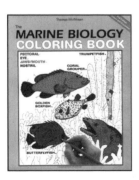

**The Marine Biology
Coloring Book**

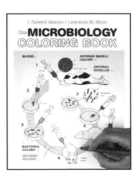

**The Microbiology
Coloring Book**

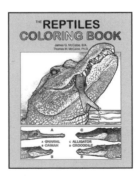

**The Reptiles
Coloring Book**

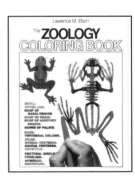

**The Zoology
Coloring Book**

For more information, visit coloringconcepts.com